Howard Pixton

*Test Pilot and
Pioneer Aviator*

The Life and Times of a Pioneer Aviator

Howard Pixton

The Biography of the first British Schneider Trophy Winner

Ghosted by
Stella
Howard Pixton's daughter

Pen & Sword
AVIATION

First published in Great Britain by
PEN & SWORD AVIATION
an imprint of
Pen and Sword Books Ltd
47 Church Street
Barnsley
South Yorkshire S70 2AS

Copyright © Stella Pixton, 2014

ISBN 978 1 47382 256 6

The right of Stella Pixton to be identified
as the author of this work has been asserted by her
in accordance with the Copyright, Designs and Patents Act 1988.

A CIP record for this book is available from the British Library.

All rights reserved. No part of this book may be reproduced or transmitted
in any form or by any means, electronic or mechanical including
photocopying, recording or by any information storage and retrieval
system, without permission from the Publisher in writing.

Printed and bound in England by
CPI Group (UK) Ltd, Croydon, CR0 4YY

Typeset in Times by CHIC GRAPHICS

Pen & Sword Books Ltd incorporates the imprints of
Pen & Sword Books Ltd incorporates the imprints of Pen & Sword
Archaeology, Atlas, Aviation, Battleground, Discovery,
Family History, History, Maritime, Military, Naval, Politics,
Railways, Select, Social History, Transport, True Crime, and
Claymore Press, Frontline Books, Leo Cooper, Praetorian Press,
Remember When, Seaforth Publishing and Wharncliffe.

For a complete list of Pen and Sword titles please contact
Pen and Sword Books Limited
47 Church Street, Barnsley, South Yorkshire, S70 2AS, England
E-mail: enquiries@pen-and-sword.co.uk
Website: www.pen-and-sword.co.uk

Contents

Foreword vii
Preface viii
Previews ix

Chapter 1   Man is Not Meant to Fly 1
Chapter 2   Flying Here to Stay..8
Chapter 3   Turning Point 15
Chapter 4   With Roe at Brooklands..20
Chapter 5   With Roe flying the Avro Biplane 54
Chapter 6   With Bristol at Brooklands 75
Chapter 7   With Bristol at Salisbury Plain 98
Chapter 8   With Bristol Overseas 111
Chapter 9   With Bristol as School Manager......................................142
Chapter 10   With Sopwith Flying the Tabloid Biplane 154
Chapter 11   With Sopwith winning the Schneider Trophy 161
Chapter 12   With Sopwith after winning the Schneider Trophy 176
Chapter 13   War 1914–1918..188

Index 226

DEDICATION
This book is dedicated to my father Howard Pixton.
British born, employed by A.V. Roe as his first test pilot,
he tested, flew and advised on many strange machines.
In those early pioneering years, during which he won
the Schneider Trophy, he risked life and limb
to advance aviation for his country.

THANKS
My thanks to my son Robert Dallas,
to my friend Frank Burlingham,
to Brigadier Henry Wilson of Pen & Sword,
for their advice and preparation of this book.

Thanks are belatedly extended to all wonderful founders of *Flight, Aero* and *Aeroplane* magazines with special thanks to C.G. Grey for his great friendship with my father, and also to the newspapers of the day, especially the *Daily Mail* under the leadership of Lord Northcliffe, for their detailed record of this period of history in which my father came to be involved.

Foreword

The name Howard Pixton is little known in most aviation circles, yet his contribution to aviation history is immense. He won the Schneider Trophy in 1914, having learnt to fly at Brooklands in 1910 at the very dawn of British aviation holding Aviation Certificate number 50.

His links with the Avro Aircraft Company also began in 1910 and he was much involved with early aircraft design, not least one of the greatest ever training machines, the Avro 504, on which my father was later to learn to fly at RAF Cranwell in 1920.

Howard Pixton taught many of the early RFC and RNAS pilots to fly, but at heart he was an engineer and did great work in both world wars as an expert on accident investigation. This was the forerunner of that work at Farnborough that now has a worldwide reputation for excellence.

His life-long love of the Isle of Man began after the First World War and his daughter Stella, who has contributed so much to this book, lives there to this day. I first met Stella when I was living on the island in the early years of this century as Lieutenant Governor; she is rightly proud of her father's place in British aviation history, and her shop, Jurby Junk, is a place not to be missed on any tour of that lovely island. Howard Pixton is buried in the nearby Jurby churchyard, formerly the church for RAF Jurby.

This is a story of the very earliest years of aviation, told by a man who was intimately involved with it all, whose passion was aviation, and who had the good fortune to work with all the great aviation pioneers in Britain. It is most appropriate that, in the centenary year of his Schneider Trophy victory, the story of Howard Pixton's early life, and his contribution to aviation, should now be told.

I.D. Macfadyen
Air Marshal
Constable & Governor
Windsor Castle

Preface

I have been asked...no, instructed to write a piece about my grandfather, the shining light of Brooklands, the aviation pioneering hero.

I didn't really know him that well. When at boarding school, and hearing of his passing, I cried, because I thought that was what was expected of me.

My 'true hero' of this book is Stella, a shining star in the firmament.

A woman who, when head of her department at the 'Central Office of Information' in the late 60s, would go to work wearing Elton John electric blue boots and an all-in-one ocelot cat suit and wonder why she was overlooked for promotion.

A woman, whose desire to have a child from every race in the world, started and stopped with me...thank God!

A woman, who has overcome her own battles in life and still has faith in humankind.

What a woman...what a unique woman.

So I would like to thank my mum, my hero, my Stella for this book, because it is through her that I have got to know my grandfather. His story is 'history' in the making...truly epic...groundbreaking...record-breaking...heart breaking, and as I read, I am at his side living every moment of his early pioneering life.

Your father would be proud of you 'Stel'...Sonny Boy is.

Robert Dallas
(Grandson)

In the early years of the last century, aviation was cool and aviators were the rock-stars of the day. These magnificent men and their somewhat less magnificent flying machines were brave, romantic and quite mad. A clique of dashing young Edwardians took up flying, and often dying, in a weird array of hazardous contraptions held together with bits of string and pieces of wood. Yet these heroes paved the way for speedy and safe international flights for the masses. One of the most accomplished yet often overlooked members of this fraternity of pioneers was Howard Pixton, winner of the 1914 Schneider Trophy, whose fascinating story as told to his daughter Stella, is the subject of this book. So many of these airmen perished during these flights and while attempting to break records, yet Howard Pixton, one of the greatest, lived until the 1970's and now his stories will live on.

Mitch Murray
(Hit songwriter, speech-writer and author)

Previews

Stella did a marvellous thing in getting her father's story down on paper some four decades ago. This fascinating account contains details of the early flying schools at Brooklands and Larkhill and of the runaway British victory in the 1914 Schneider Trophy race at Monaco, that are probably not recorded anywhere else, and but for Stella these would have been lost forever. This is vintage stuff... a cracking good read.
Captain Bob Beese *(Ex Fleet Air Arm and Dan-Air Pilot).*

The bravery and pioneering spirit of these men in this book is tangible. Their unstoppable urge to fly, moved aviation from primitive beginnings of a few hardies to become an industry in a few short years. Howard Pixton's contribution in shifting the lead from France to Britain is well described. By winning the Schneider Trophy Howard Pixton put Great Britain on the aviation map. There is comradeship, sadness, high adventure and excitement amongst an array of international characters... you won't put it down.
Honorary Colonel Charles Wilson. *(Holder of the Presidential Sword and Chairman, Royal Aeronautical Society, Isle of Man Branch).*

The story of the life and times and achievements of Howard Pixton, Britain's greatest pioneer aviator and winner of the world famous Schneider Trophy Air Race in 1914... A remarkable man, who came to the Isle of Man to retire in 1932, raised a family and when he died in 1972 was buried appropriately at Jurby Church overlooking the former Royal Air Force aerodrome. In the words of Shakespeare 'some men are born great, some achieve greatness and some have greatness thrust upon them'... Howard Pixton achieved greatness and deserves recognition.
David Cannan *(former MHK, Treasury Minister and Speaker of the House of Keys, Isle of Man Government).*

The story of Howard Pixton should be an inspiration for aeronautical engineers both now and in the future to accept that there are no boundaries between earth and the heights of aviation.

Maxine Cannon *(General Manager, Stamps and Coins, Isle of Man Post Office)*
It was the least powerful aircraft in the race barring one other with a similarly puny engine. And not only was it inferior on horsepower, it was also the smallest aircraft in the contest. Yet these apparent drawbacks were to be turned to advantage by a pilot who, with a deft touch and outstanding airmanship, not only won the world's most famous air race but also claimed supremacy in aeronautics for Great Britain for the very first time. Captain Howard Pixton, for far too long the 'unsung hero', is finally at the 100th anniversary of that glorious Schneider Trophy victory, elevated to his rightful place in the pantheon of aviation's all-time greats.
Nigel Macknight *(Aerospace author)*

I am convinced that Howard with his masterly handling of the Tabloid put confidence into British aircraft designers and manufacturers at a time when it was most needed, just before the outbreak of war. I honestly believe that Howard did more for British Aviation than any other pre 1914-18 pilot, or any other Aviation enthusiast of that time.
Squadron Leader Tom Gleave *(WW2 Fighter Pilot and No1 Guinea Pig for early plastic surgery who read a pre-edited version of this book).*

Brooklands June 1910
The World Famous Automobile Racecourse

Brooklands, showing the famous experimental flying centre within its 2.8 mile racing circuit at the time when Howard Pixton arrives to learn to fly with A.V. Roe. The plan shows its early flying huts that were numbered 1-18, with Roe's hut No. 14 marked, and the River Wey still with its loops. During 1907, when the course was being built, part of the sewage farm had to be contained within the course due to the expense involved in removing it.

Brooklands, where Howard spent the happiest days of his life...

BROOKLANDS RACECOURSE AND FLYING GROUND 1910

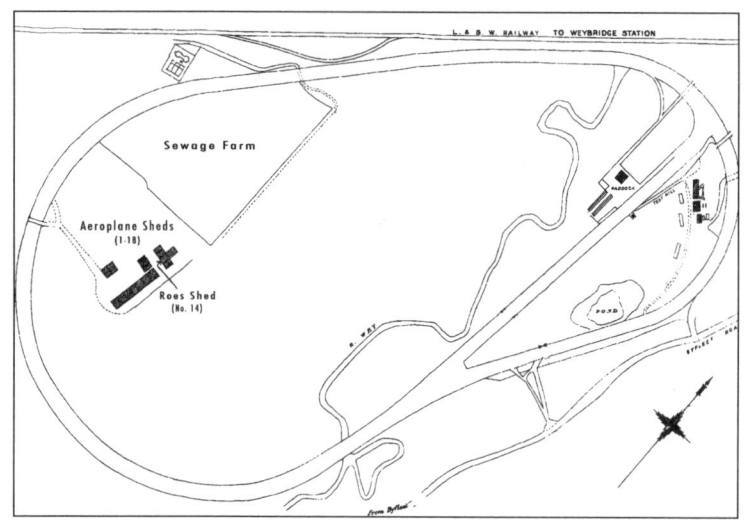

Images of Brooklands early flying days

CHAPTER ONE

Man is Not Meant to Fly

I was there... my full name is Cecil Howard Pixton, but I have always been known as Howard.

I was born during Queen Victoria's reign on 14 December 1885, in 'The Year of The Motor Car' when, with the advent of petrol refinement Benz and Daimler, both German engineers working separately, were designing petrol-driven engines for their 3 and 4 wheeler vehicles... then during 1886 when I wasn't yet one-year-old, the great science-fiction writer Jules Verne, was putting forward his ideas in a novel of his fictional travels in a heavier-than-air vessel, *his ship in the air*, as being far superior than any lighter-than-air balloon flight...

> 'I wanted the air to be a solid support to me, and it is... I saw that to struggle against the wind I must be stronger than the wind, and I am... Air is what I wanted, that was all. Air surrounds me as water surrounds the submarine boat, and in it the propeller acts like the screw of a steamer... That is how I solved the problem of aviation...'

We lived in comparative comfort in quite a large house at Palatine Road in the residential district of West Didsbury, having two maids, a cook and gardener. There were four boys in the family, Dudley, Aubrey, Bert and myself the youngest. My father, a well-known stockbroker, had his business premises in Half Moon Street, St Anne's Square, and while I was very young, formed a partnership, the firm then being known as Messrs Pixton & Coppock. Each year we went to the Isle of Man for a holiday, staying five or six weeks at the Bay Hotel, Port Erin. I spent a lot of my time in the open-air swimming pool there, taking home medals on winning various speed swimming competitions. We also spent a week or so at Llandudno where it was very quiet and suited father whenever he wanted to get away from his business for a few days. When of age my father sent me to a Manchester Grammar School as a day pupil where we had a fine headmaster who amicably referred to my youthful declarations as 'Pixtonian Words of Wisdom'. However, a retired sergeant who came to the school from time to time and gave us wooden guns for drill instruction, was not so kindly disposed towards me, and would bellow, '*Pin-xton. You Stoo-pid lad!*'

Horse cabs were the usual form of transport and, in common with most youngsters, I used to ride on the cab's back axle, but more than often children playing in the streets shouted up to the coachman, 'Whip behind, Guv'nor.' And the driver understood he had unwanted passengers and would crack his whip smartly over his head so that it caught our faces and limbs. Sometimes because of this joyriding, spikes were fitted onto the axle, thereby depriving us of our free ride and fun of daring, but with the discovery of petrol there was a new vehicle, the motor car, first known as the horseless carriage. Horses were very frightened by them and easily went out of control when one came near, but seldom was one seen. The few passing through Manchester had a man, usually a servant or friend, walking in front with a red flag to ensure a speed limit of 4 miles per hour was kept for public safety. It was illegal, otherwise, to drive, but these strange vehicles attracted us when on the rare occasions they appeared and we would run beside them for miles until exhausted. The men with the red flags were supposed to stay in front, but if the Police were not in sight and the road was quiet, they invariably jumped aboard. Unfortunately our Government held us back in the new industry with its absurd restrictions so that we were far behind European countries and America, not only in our attitude towards the motor car, but in our engineering knowledge. The few seen in Britain were imported chiefly from France which led in their manufacture. Then the red flag days were over. From 14 November 1896, motor enthusiasts were at liberty to race through the streets up to 12 miles per hour unaccompanied. Headline news – 'Emancipation Day for Motorists'.

Aubrey gave me a bicycle made with thick steel tubes and pneumatic tyres stuck to the rims of the wheels, and one day while out cycling I saw for the first time a car travelling at the new speed of 12 miles per hour. I'd not seen a car for three years since the 4-mile limit had been lifted as so few people owned one. I jumped onto my bicycle to follow it. What a difference in speed. It was coming from the 1899 Agriculture Show at Trafford Park where it had been on display, and as I followed closely behind, doors all along the streets were flung open and people rushed out to see it pass by. Although cars were still a novelty, the public expressed unbelievable prejudice towards them. I loved them and hoped to have one of my own when I was old enough, but the Press gave their opinions and published readers' complaints, '*Ban them from the roads... noisy, dangerous playthings for the rich*'.

Petrol was also being put to a different use. A number of balloon flights had been made in various parts of the world, but when I was 13-years-old a very practical, elongated one, steerable by the use of power derived from petrol, was headline news. It had been constructed by Santos-Dumont, a resident in Paris and son of a wealthy Brazilian. In the following years Santos-Dumont used his dirigible as an everyday means of transport. Parisians waited to see him come home by air to his apartment in the Champs Élysées, and watched his servant catch the guide rope to bring him to earth.

Just before I left school in 1901, I was the very proud possessor of a 2¾-horsepower Ariel Tricycle. I thereby became one of the first owners of a motorised vehicle in Manchester. It had no number plates but when the Motor Act was passed and car registration became law, I had to fix a plate, N78, onto it. Father and Coppock dissolved their partnership at the beginning of the century, Aubrey joined the company then registered as J.S. Pixton & Son, but not long afterwards father told us the state of his affairs. He was bankrupt. The news was unexpected and came as a terrible blow. Although I asked what happened, I never really understood what had actually gone wrong. We left Didsbury, had a short stay at the Bowdon Hydro and, still remaining in Manchester, moved into lodgings at 31 Claremont Road, Alexandra Park. Father became the manager to the Manchester Safe Deposit Company, and I was apprenticed to the Industrial Engineering Company at Newton, Hyde, a company that made a variety of things. During my short stay with them they built the first motor-horse, a steam truck on wheels with a detachable engine cab to pull it. It was a revolutionary idea, but the vehicle did not prove efficient on steep hills as the wheels spun round due to insufficient weight and lack of gripping power in the iron tyres.

Motor vehicles certainly fascinated me. What fascinated me even more was the new vehicle of the air, the Flying Machine. Up to now no one had managed to fly a heavier-than-air machine, so when the world's first flight was announced, I was filled with avid interest. For hundreds of years man had tried to fly. It was December 1903 when two Americans, Wilbur and Orville Wright raised their flying machine, *The Flyer*, off the ground running it along a rail to gather speed and using a petrol-driven engine working two propellers behind the wings. It had skids, no wheels. What an achievement! A scientist some centuries ago had concluded that man could never hope to succeed to fly, '*God would prevent such a revolution in human affairs.*' More recently, it was thought that an object heavier-than-air would never leave the ground unless lifted by gas and that balloon flight was the only means of air travel, but the brothers had flown between 200 and 300yds.

It took the brothers another year to fly three miles and they were the only two who knew how to fly by the end of 1905. By then they'd flown 24 miles in a single flight. It was incredible! Two years had passed and still no one else knew the secret of flight. France had a few designers with machines that could not fly and Britain had nothing except balloons, kites and the odd man or two who dreamt of flying. But people all over the world doubted the brothers' achievements, believing their claims extravagant. '*We've not seen them... Why don't they show us?.. No one can fly... Man is not meant to fly!*'

I served my three-year apprenticeship with the Industrial Engineering Company, and in 1905 joined another firm, the Simplex Engineering Company at Trafford Park, makers of four-spindle automatic lathes and machine tools, and was employed in the drawing office as a machine-tool draughtsman. It was a good firm to work for and while there I bought myself a 4½-horsepower Minerva motorbike, a great improvement on the Ariel Tricycle. I'd been at the company for just over a year when, in 1906, three years after the first flight, Santos-Dumont was in the news again. The well-known balloonist had flown 200yds on a flying machine in France, so becoming the second man to fly, counting the Wright brothers as one. With this latest news the French were jubilant. Of his achievement, they exclaimed 'What a triumph. A month ago Santos flew 20yds. A fortnight ago he flew 70. Yesterday he flew still further. The air is truly conquered. Santos has flown. Everybody will fly.'

Next day the founder of the *Daily Mail*, Lord Northcliffe, offered £10,000 for the first man to fly between London and Manchester. Some admired him, 'A great enthusiast'. I was just amazed by the enormity of the sum, £10,000 in 1906 was quite a fortune, but as most people still believed flight was not possible, such an offer was met with ridicule. '*It can't be done... It'll never be possible... It'll be a miracle to do a mile, never mind 183 miles... Publicity exploit of the Daily Mail!*' Their offer was ridiculed by a rival paper, *The Star*, who offered £10,000,000 for a lesser flight.

> 'A morning paper makes the trivial offer of £10,000 to the first aeroplane that flies from London to Manchester. Our own offer of £10,000,000 to the flying machine of any description whatsoever that flies five miles from London and back to the point of departure still holds good. One offer is as safe as the other.'

The proprietors were sticking their necks out a bit, but still the Wright brothers' flights were not believed. Had they come to England, *The Star* would have been placed in a very embarrassing position as the brothers could have easily flown the desired distance without much difficulty. Then *Punch* entertained its readers with an offer of £30,000 in three sums of £10,000 for three improbable tasks:

'£10,000 to the first aeronaut who succeeds in flying to Mars and back within a week... £10,000 to the first person who succeeds in penetrating the centre of the Earth in a fortnight... £10,000 to the first person who succeeds in swimming from Fishguard, Wales to Sandy Hook, USA before the end of 1909.'

Following the *Daily Mail*'s lead, there came other serious offers, but none as great as theirs. Lord Montagu, editor of *The Car*, backed their offer with £1,000 or £5 for each mile flown. In addition, Lord Montagu, as President of the Brooklands Automobile Racing Club, announced that he would pay £2,000 to the first man who flew around their newly built track at Brooklands in Surrey, if accomplished before or just after the end of 1907. Thomas Holt, director of *The Graphic*, also offered £1,000 to the first to fly a mile at Brooklands with a passenger. Just trying to get a machine off the ground was difficult enough, never mind with a passenger.

Hoping to win the Brooklands prizes was A.V. Roe, a Manchester man like myself, and soon after the announcements he went to the race track with a biplane he had built, which had one propeller behind the wings. There, at the track, motorists offered to tow him to give his machine lift at the start of the attempted flights, but all was of little avail. For most of that year after the racing season was over, A.V. Roe was alone at Brooklands, but as much as he tried, the little 9-horsepower engine he was using had not the power to lift his machine sufficiently off the ground. He had been waiting for a French engine, a 24-horsepower Antoinette, with which he would have had a better chance, but it did not arrive in time. The prizes were withdrawn and Roe left Brooklands at the beginning of the 1908 racing season.

While a handful of British experimenters were trying to fly, a third man lifted a machine over 200yds. He was Louis Blériot, a Frenchman, and he did it in September 1907. He'd not copied the proven Wright machine, but had worked on entirely new principles regardless of peoples' opinions, and had produced a monoplane which met with no encouragement. '*A flying machine with only one pair of wings could never fly... It'll not have enough lift.*' But fly he did. What was so important about Blériot's success was that the world now had two basic designs to copy, the biplane and the monoplane. Not only that, but even more radical was the fact that he'd placed the propeller on the nose of the machine. Thus the terms 'tractor' and 'pusher' originated to distinguish the type of machine, depending on the position of the propeller, in front or behind the wings.

A couple of months later two more men flew over 200yds; Henry Farman and Delagrange had got a Voisin Box Kite to fly. It can be said that Farman was the first Briton to fly, but his successes were not considered British as his claim to British nationality was practically discounted. His father was English, the Paris correspondent of *The Standard*, his mother French and I believe Farman himself was born in Paris or had lived most of his life there.

A long time had elapsed since the Wright brothers were in the news. As no one had heard much about them since 1905, further doubts to the validity of their claims arose. No one had yet flown in Britain, Blériot and Farman were managing short flights, and the Wright brothers were almost forgotten when suddenly they showed themselves to the world. They arrived in France in mid-1908 and flew, to the amazement of French designers, astonishing non-stop periods of an hour or more, covering 30 to 40 miles in a single flight launched from a derrick and rail. Why had they not come forward earlier? Why had they not reaped the several thousands of pounds of French and British awards on offer? Would they now try for the *Daily Mail* £10,000 for the London to Manchester flight, still open to anyone of any nationality? The reason why they went to France was to discuss production with the French Government since they were not backed by their own government. They'd already approached Britain but no agreement was reached, as the War Office had not been interested in their demanding propositions.

Shortly after the Wright brothers had flown those fantastic distances in France, one man in Britain flew over 200yds. This man was Cody, an American employed since the beginning of the century at the Government Establishment, Farnborough. He'd been engaged for his kiting knowledge, and had made several man-lifting kites for the Army, from which followed the construction of a flying machine financed by Government money. So at last Britain could fly.

Cody's flight, accomplished during the autumn of 1908 was a great British achievement, but compared with the Wright brothers' and French standards, it rated as an insignificant flight. The machine resembled the Wright brothers' pusher biplane but what Britain needed was more men interested in flight to catch up with the great lead of America and France. It had been just the same with the motorcar, America and France ahead of us! Roe worked quietly on his second machine, a triplane, having three pairs of wings and propeller in front. He had risen, but for no appreciable height or distance. Meanwhile, there'd been reports of a marvellous flying machine being made in secret in Glen Tilt, Scotland. Dunne, a friend of the famous writer H.G. Wells was experimenting with a machine that had its wings swept backwards in a wide V-shape. A great interest was taken in it but these early experiments were not successful, for had it flown everyone would have known about it.

During my employment at the engineering companies, I'd spent six years, from 1902 to 1908, as a day and evening student at the School of Engineering, 78 Cross Street, Manchester, and passed in the higher stages of Machine Drawing, Applied Mechanics and Practical Mathematics in various examinations. J. Firth, the principal, wrote in a recommendation, 'He has an aptitude for the Engineering Profession generally and a liking for, and inclination towards motor-work in particular.' I had, by now, been three years at the Simplex Engineering Company, and was assistant to the chief draughtsman, but I decided to change from my white-collar position and put on overalls in order to enter into the practical side of engineering.

I found a situation as a mechanic with the British Engineering Company of Leek, manufacturers of steam and gas engines, dynamos and motors, and left Manchester. On arriving at the small town I looked for modest accommodation and happened to go along Barngate Street. The rows of houses all looked alike, each with a front door and window alternating down the street. I went from one house to another, knocking on all doors until I found an elderly lady who had a spare room. 'How much?' 'Full board, fourteen shillings.' That was very fair. I took it. She had three rooms to let, the other two were occupied by a couple of Mormons, very nice fellows in their fifties who went around the town in search of new followers. They did not stay long in Leek as I imagine the locals were not easy to convert and two workmen took their place, staying with us while they installed a radically new type of gas engine at the local electric station, made by Mather and Platt of Manchester. The front room was the common room for guests, which also served as the dining room, but I'd not been there long when I was wandering into the kitchen and taking my meals with the landlady. I learnt that her husband had died some time ago and her only son was a chemist in London.

The firm I'd joined was owned by people called Sutcliffe and I did all sorts of jobs for them. They had a horse and trap and employed a coachman, but had just replaced them with a 20-horsepower Beeston Humber car, the more expensive of the two types, the other being the Coventry Humber. They were exceptionally nice people and during the period I was in Leek I was extremely happy, the work was pleasant, the people of Leek friendly and the surrounding countryside very beautiful.

But I was still extremely interested in flight...

CHAPTER TWO

Flying Here to Stay

A publication, *Flight* founded by Stanley Spooner at the beginning of 1909, was to be a magazine entirely devoted to the new subject of the flying machine. Up to this time the *Aeronautical Journal* and motor magazines had carried the news on flying matters. I bought *Flight* each week to read of the latest developments but it was full of French news. There appeared, however, short notes on lectures taking place in Britain, the formation of flying clubs up and down the country whose members contented themselves with ballooning, kiting, model-making and model flying, none of which particularly interested me. Some readers thought that machines should be called 'aerodromes', a word meaning 'air runner', and then there was an article on what were thought to be the best machines, but no mention of Blériot.

> 'The most successful types of flying machine or aerodrome at present in existence are those constructed by the Brothers Orville and Wilbur Wright of the USA and by the Voisin Frères of Billancourt, Seine.'

Four months later another publication, *The Aero*, was launched. It was obvious that the aeroplane was here to stay, although airships were considered vital and several countries were going ahead building them, especially Germany who had successfully built a number of Zeppelins.

There was not much doing in Britain except that the Society of Motor Manufacturers and the Aero Club were arranging for the First British Aero Show to take place in London, at Olympia, organised on the lines of a recent one in Paris. When the show did take place during March 1909, the Voisin was the only practical machine on display. In fact, only eleven machines, seven of which were French, were shown beside balloons, kites, models and motor engines. There was no Wright flyer, no Blériot, nor the Antoinette, a beautiful French machine making a mark for itself. We'd managed to conjure up four of our own: an unfinished Short, a Howard Wright, a small Weiss, which was an amateur design that had interested Handley Page, and another unfinished machine, a Lamplough which possibly would never fly. Prices ranged from £500 to £1,400.

I learnt that a flying field was nearing completion at Shellbeach, Isle of Sheppey, which the Short brothers had prepared in conjunction with the Aero Club. They'd converted a thirteenth century house into a club for members, and had started a works there building Wrights under licence as well as their own machines which were nothing less than modified Wright pusher biplanes. However, the Shorts Company was the first to be set up in Britain in a practical manner. Several Englishmen had gone to France to buy machines since none were available in Britain, as did Moore-Brabazon who returned with a French Voisin Box Kite which he flew at Shellbeach during the early months of 1909. The Aero Club looked upon Moore-Brabazon as the first person to fly in England. A.V. Roe thought he was even before Cody.

There was a discussion at the Mansion House between the Lord Mayor of London, prominent people and reporters where it was urged that the British Government should substantially finance and support the development of the airship.

> 'We are an unprepared Nation... We must, at all costs, have airships... The only airship to hang over the Bank of England or the Mansion House must be flying the Union Jack.'

People could understand balloons and airships, lighter-than-air flight, but the aeroplane was still a subject for ridicule. Nevertheless, Lord Northcliffe offered further large sums of money for the first to accomplish specific flights. Not only the prize of £10,000 for the London to Manchester flight was still open, but another of £1,000 was offered to the first man to cross the English Channel. Though smaller, it was still a huge sum of money. And in July 1909 the Channel was crossed, the first of the *Daily Mail*'s awards was claimed and the whole world knew of it.

It all started when a Frenchman, Hubert Latham, took up the challenge. Latham had learnt to fly an Antoinette with exceptional success. Like Farman he was born in France but was not considered British, although his father was English. He decided to cross the Channel from Sangatte and on the selected day, the weather seemed favourable and Latham set off. He didn't get very far. A few miles off the French coast his engine failed. He dropped into the sea and sat afloat in his plane until a dinghy picked him up. A piece of loose metal was later found in the engine. He would try again. However Blériot, now in his mid-30s and a more experienced flyer, set off before him. Blériot had just flown a magnificent 25 miles across country and believed the Channel crossing was within his reach. This was so. He left the small village of Barraques, near Sangatte, in his latest machine, *The Blériot XI*, and forty minutes later saw his friend waving a tricolour on the Dover cliffs. He landed safely in North Fall Meadow and crowds swamped him and his machine as they rushed to congratulate him.

Blériot had spent an estimated £30,000 over the last few years designing and trying to produce a machine that could fly, but now he had a thoroughly tested machine that had carried him across the English Channel, man's first sea crossing in an aeroplane. A wonderful flight! France became the leader of the world! Lord Northcliffe met Blériot at a packed Victoria Station and the aviator was driven through cheering throngs of people lining the London streets to the Savoy Hotel. During luncheon he was presented with the well-earned £1,000 reward and was surrounded by reporters wanting to know what his flight had been like. 'Out of sight of land, I felt as though I was not moving at all, having nothing on which to fix my eyes for judgement of speed. Visibility was not very good, and I had a little trouble with the engine overheating, but the drizzle helped to keep it cool.' Madame Blériot, mother of five children, was with her husband. She'd known him to crash on many occasions, and seemed pleased when he announced, 'I will be giving up flying soon and employing pilots.'

Blériot and Latham were friends. By radio message, a new means of communication, Blériot said that if he could cross the Channel on that same day he would share the prize money, but the wind had risen and Latham could not manage it. He tried shortly afterwards but his engine failed a second time and once more he dropped in the sea. Blériot's monoplane was put in Selfridges, the new London store in Oxford Street recently opened by the American, Mr Selfridge. For the next four days thousands of people came to see it. I wondered why the Wright brothers had not crossed the Channel.

By now I was fired with enthusiasm. My one consuming ambition was to fly! But how could I possibly get into the flying world? What could I do? What chance had I with a wage of 25 shillings a week when machines cost anything more or less between £500 and £1,000? My father could not help and I couldn't buy one of my own. Besides, there still weren't any to be had in England, and even if I could go to France I would be lucky to find one there as the few in production would have long been ordered. Only the more well-off men were flying. I saved every penny with one object in view, to fly. Although most people, like myself, had not seen a machine, now that Blériot had flown the Channel an interest was being taken in flight. The impossible had been accomplished, but many still had doubts as to the future of the aeroplane. *'Flying will never be any real use... Flying can never amount to anything serious... A hobby for those who can afford it.'* Then newspapers were receiving letters from another section of the public who expressed their dread of the future when aeroplanes might fill the skies, believing that small articles might drop from them, and called for protection. *'The aeroplane should be taxed out of existence!'* Far from that, the Chancellor of the Exchequer did his best to help the constructor by exempting petrol used for aeroplanes from tax.

In August,1909 just after Blériot's Channel flight, the first large Flying Meeting was held at Reims. France, Britain, America, Austria and Italy were represented, Cockburn was the only British competitor. Throughout the eight days of events, it was wet and dull, which did not help matters. Although there were quite a large number of machines present – nearly forty with about twenty-eight owners – there were only ten different makes in all, the better-known ones being the Wright, Blériot, Antoinette and a Henry Farman Box Kite. The first machine brought out got stuck in the mud. After its brave owner tried desperately hard to fly it, it was towed back to its hangar. As only about a dozen machines left the ground, a lot of money exchanged hands as those who could not fly had to forfeit 1,000 Francs per machine.

At the meeting, the Gordon Bennett Aeroplane contest was inaugurated. Gordon Bennett, proprietor of the *New York Herald*, was the man behind the world famous Gordon Bennett Balloon contests. Glenn Curtiss, a fellow American, took the trophy by being the fastest around two laps of $12^{1}/_{2}$ miles and the honour of holding the next contest went to America as the result of the win.

In September, I left Sutcliffe to work at the only garage in Leek to gain further mechanical experience. The owner, a very pleasant man called Magnier, possessed two Darracq French cars which had the gear change on the steering column. Apart from the necessary mechanical jobs which needed doing, I drove people out in these cars on a hire-service arrangement as part of my job. Mr Magnier knew of the interest I had in flying and one day, after some people had called, he told me that they were looking for someone to drive them to Germany, and on the way intended to stop at the Aeronautical Exhibition in Frankfurt. It would be quite an undertaking, but would I take it on? I was only too delighted to do so. They had a Coventry Humber for the trip. It was my first time abroad. All went well we arrived in Holland, drove through France and uneventfully into Germany. There, at the Frankfurt Show I saw my first aeroplane, a Voisin, the only one on view. I was thrilled. I thought I might see more than one but the interest lay in airships. The catalogue I think was about 300 pages thick, showing countless models of airships, airship hangars, anti-airship ammunition displays, and models of aeroplanes, some working with catapult propulsion, others just the designs and hopes of machines that might fly if blown up to size. But the airship was Germany's pride and joy.

Enthusiasm shown in Zeppelins by the ordinary person was enormous. Shops sold pictures of the elderly inventor, Count Zeppelin, ornaments, jewellery, handkerchiefs, scarves, cloths, spoons, were all decorated with either his portrait or airship, and even his autograph. I bought a novel penknife in the shape of a Zeppelin and several amusing postcards depicting the excited public as a Zeppelin passed overhead, each headed with *'Zeppelin Kommt.'* We drove on to Düsseldorf. I was pleasantly surprised to see a Zeppelin in flight over the town and it landed on the parade ground right in front of me on being forced to descend because of a high wind. As the huge ship came down, ropes were flung from the gondola when it was about 40ft off the ground and were caught by soldiers who pulled it down until passengers were able to step out. It was quite a sight!

At the end of October, we returned to Leek. In our absence, Louis Paulhan had been brought from France at the great fee of £1,000, to show the British people some flying, the spot chosen was Brooklands Race Course. Hundreds of paying spectators turned up to watch him. At last the British were waking up to flying! It became obvious to Major Lindsay Lloyd, clerk of the course, that flying was a potential crowd puller, so he arranged for the centre area of the ground to be cleared and flattened, then advertised it as a flying field where sheds could be rented. The response from experimenters was overwhelming, and he was kept busy for months meeting demands and preparing Brooklands as a flying centre. I learnt also that in our absence Cody had flown a mile, the first man in Britain to do so and about the ninth in the world.

Another of the *Daily Mail* flying prizes had also been claimed: £1,000 for the first Englishman to fly a circular mile on an all-British aeroplane. It had been taken by Moore-Brabazon on a Short plane at Shellbeach during October 1909, while watched by Aero Club officials, using a derrick and rail, a continuation of the Wright Brothers' idea of aiding their flight. Other machines like Cody's and Roe's could lift of their own accord.

The news of Moore-Brabazon's flight and that he was the first Englishman to fly in Britain, brought a letter to the flight offices from County Roscommon, Ireland. 'You refer to Mr Moore-Brabazon as the first Englishman to fly in a heavier-than-air machine. Over here we call him an Irishman.'

The first flying meetings in Britain had also taken place, one at Doncaster and the other at Blackpool. At Doncaster Cody was invited to fly by its promoters for a fee of £2,000 and there he signed the documents by which he became a naturalized British citizen. A.V. Roe had attended the meeting, but did no more than a few hops on his new triplane. Cody, without a doubt, was the most well-known flying man in Britain, but for some time now he'd been dogged by bad luck. He'd notified the Aero Club that he would attempt the circular mile, but Moore-Brabazon had beaten him to it. Moore-Brabazon had been long associated with the club. Cody had tried to fly from Liverpool to Manchester, which would have resulted in winning £1,000 had he succeeded, an offer made by Sir William Hartley of the Hartley Jam Company, but he had to abandon the flight because of fog, then the offer was withdrawn. He'd also intended to try the *Daily Mail*'s £10,000 London to Manchester, and had been all set to leave when his engine refused to function, but set-backs and frustrations were to be expected. Flying was still very much in its infancy.

Making and buying machines was a personal thing in 1909 but slowly an aeroplane industry was developing in Britain. The Short factory was well underway at Shellbeach, the Handley Page Company at Woolwich was advertising aeroplanes for sale, the famous Humber motorcar works at Coventry were producing the Blériot under licence and one of their own design looking very much like the Blériot, but the Wright brothers were causing problems in the aviation world. They'd started legal action against various firms and pilots in France whom they considered had infringed their patents, but I don't think they, nor their licensees, although winning their cases, had succeeded in receiving much sympathy. Even Cody had had a letter from them, but the matter was dropped when he pointed out that he'd used wing warping on kites long before their patent had been registered. But the brothers continued to raise objections.

> 'When the so-called marvellous advance of French aviation in the past year has consisted chiefly in copying more and more closely the main features on which our patent is based, then we think the advancement of the art will not suffer greatly if these imitators are compelled to do some real experimenting and inventing for themselves, instead of step-by-step bringing their method of lateral balance to an exact copy of ours.'

The fact remained that France was leading the world in flight with Britain trying desperately hard to catch up while the Wright brothers were not doing much, being mainly concerned with legal proceedings against everyone, even their fellow American, Glenn Curtiss, whose counsel ridiculed the idea that the Wrights had any special monopoly. Curtiss himself said, 'The proprietors of the Wright patents think no one can make aeroplanes without them, and they're about the only people who think so.'

And so ended 1909, but not without various opinions expressed by prominent people in France regarding the future of the aeroplane.

> *'I believe that the aeroplane will exist only for sport, and will be reserved in consequence for a very small number of users... Social life will be transformed by the flyer... Flight will have the effect of virtually turning modern houses upside down. The sixth floor will be considered best, being in closest proximity to the aerial way... In one hundred years hence, the great cities will practically be uninhabited except for business purposes during the daytime. The motorcar will have gone completely out of fashion, but the bicycle in a new form, will be once more in favour, for a sort of flying bicycle will be invented which will enable the rider to soar in midair... One can not go against Nature, and Nature has only created the birds to make us lift our eyes towards the Infinite, and to cause us to think that it is the domain of souls and spirit invisible.'*

CHAPTER THREE

Turning Point

The Aeronautical Society had been founded many years before I was born in 1866 and in 1909 the Royal Aero Club was formed. The Aeronautical Society therefore became more a meeting place for scientific discussion with the Aero Club being devoted to the practical side of flying arranging meetings, licensing and noting records established in Britain. Following in the path of the Aero Club of France, the club started issuing flying licences in March 1910 and establishing rules by which one could gain them. The first four issued were to Moore-Brabazon No 1, Rolls, Rawlinson and Grace, No's 2, 3, and 4.

I was now 24-years-old and had become very determined to fly, but should I take a chance and leave my job to look around hoping for some lucky break? I hesitated, as I'd hesitated throughout 1909, not knowing quite what to do or how to go about it in a reasonable fashion, as my chances were very slim. Many men like myself were eager to enter this new field of flight and the flight magazines were constantly receiving letters asking for advice. *The Aero* confessed that the most tempting answer to those wanting to fly without capital was that of *Punch*'s classic advice to those about to marry, '*Don't*'. I was no further forward in my ambition to fly, except for a few savings, when the greatest of all the *Daily Mail*'s prizes was claimed. *It was the turning point in my life*!

It started one Saturday morning at the end of April 1910, when Claude Grahame-White, a man of thirty-one who'd only just learnt to fly in France, was ready to try the formidable task of flying from London to Manchester. Harold Perrin, Secretary to the Royal Aero Club, with a red flag in his hand at Wormwood Scrubs, waited for him. He came into view on a Farman Box Kite, flying in from Park Royal. Perrin dropped his flag, the first flight to Manchester was under way. Perrin got into a motorcar to follow as the official observer but driving from London to Manchester alone was quite an adventure in itself. He, Grahame-White's mother, *Daily Mail* press men, mechanics and other interested parties all followed northwards as Grahame-

White followed the railway lines as his guide, parts of which had been whitewashed in preparation for this flight. Early morning risers cheered him on route and after flying for two hours, he landed at Rugby to fill the petrol tanks and was off again. Success appeared probable but a wind rose, rocking and swaying, he found it difficult to manage the controls and was forced down with a spluttering engine near Hademore railway crossing, not far from Lichfield, 113 miles and three hours from London with only 70 miles to go. The *Daily Mail* described his journey. '*He rushed through space above towns, villages, hills and valleys.*'

When the news reached Leek, I dashed off, leaving clouds of dust behind me in my haste as I drove some people to the scene. On reaching the field where the Farman stood, I separated myself from my companions, eased and pushed my way through the huge crowd gathered there to get as near as possible to the machine. It was the second aeroplane I'd ever seen and the first that had been in the air only a few hours previously. Again I was delighted just to see an aeroplane. I'd only seen photographs of the Farman Box Kite, but as I looked closer at the real thing, delighted though I was, I felt a little disappointed. This Farman, *The Henry Farman III*, fell short of my own ideas of good design. It was a mass of wires, like a birdcage, within which was an engine, the seat was just a few slats and the pilot sat fully exposed to the weather and obviously had to hold fast to a strut with one hand to prevent himself from slipping off and to the control column with the other. The Blériot Cross-Channel machine was by far superior. Even so, I wouldn't have minded trying my hand at flying it. Two men came towards us, one was Grahame-White, clean-shaven, young and wearing a khaki flying suit, and the other with a black beard was Henry Farman, the Anglo-French designer and friend of Grahame-White. As the crowd stood back to make way for the two men, I almost rushed forward with the wild idea of asking Farman if I could join him, and fly for him. What an opportunity! But I hung back. I just couldn't bring myself to such a confrontation. I spent all afternoon at Lichfield, but the weather conditions did not improve as the crowds waited patiently. Then it was announced that a check would be made at 2am when it was hoped Grahame-White would be able to fly the remaining 70 miles after getting some sleep. As the flight had to be done within twenty-four hours, he must be in Manchester before 5am, but luck was not with him. During the night the wind increased and early Sunday morning the Box Kite was discovered blown over and upside down. The mechanics had failed to peg it down securely. Stays were broken, the canvas wings were pierced, and overall damage was so great that Grahame-White had to return to London with his Farman packed on a truck for major repairs. I returned to Leek.

The *Daily Mail* made a great splash of Grahame-White's flight, but meanwhile the rather jolly Frenchman, Louis Paulhan, a far more experienced flyer than Grahame-White, arrived in London with another Farman Box Kite. He too accepted the challenge for the £10,000 award and in a field at Hendon, mechanics were busy assembling his machine. Once again Perrin was notified, Paulhan was ready to start. He flew in to Hampstead and headed for Manchester, followed by a train running at a reduced speed with Madame Paulhan and friends aboard to egg him on his way, but back in London, Grahame-White heard of Paulhan's departure and set off immediately on his repaired machine. Again he was cheered by the crowds as he left for the second time, but darkness brought them both down, Paulhan at Trent Valley station near Lichfield, Grahame-White at Roade, roughly 60 miles behind. Grahame-White left before dawn. It looked like being a neck and neck finish, but he was forced down once again by wind and a faulty engine at Polesworth near to where he'd come down before, an unfortunate coincidence for him, and only 10 miles from where Paulhan was spending the night.

Paulhan flew uneventfully to Manchester, landing in a field at Didsbury less than a mile from my birthplace. Thousands rushed to the scene shortly after 5.30am as soon as they heard of his arrival, and smiling broadly, waving heartily, he claimed the £10,000. He had flown from London to Manchester in twelve hours, stopping only once. Another impossible flight had been flown...! *The French had done it again!* They were quite a year ahead of us, and Paulhan's flight, which went off without a hitch, was yet another example of how much we were lagging behind them in aviation. It was being said, *'We, who have led nations in the development of civilisation and progress in Science, Art and Commerce, are now content to sit at the feet of our rivals'*.

Although Grahame-White had not been successful, the publicity he received from the *Daily Mail* made him now the most famous of all British flyers, and his flight before dawn went down in the Aero Club's Record Book as the first night flight in Britain. A subscription list was also opened by the club on his behalf which resulted in nearly £2,000. I thought that something might have been erected to commemorate this historic occasion at Didsbury where Paulhan landed, especially since there'd been a record of Blériot's Channel flight at Dover in the form of a stone outline of his monoplane. Neither plaque nor any such record was made either at the start or finish of the London to Manchester flight to show that it had taken place. A pity, since this was the first real bit of flying that had occurred in Britain. As for me, I was prompted into action.

Throughout the rest of April and May 1910 I must have written to practically everyone connected with flying in Britain and the replies should have dashed my hopes completely. I discovered there were no flying schools established in Britain, and the usual price asked for tuition on someone else's machine was £100. My total resources amounted to only £60. Fortunately one of my letters was to the Avroplane Company. I'd seen their advertisement in *Flight* which read:

> 'Avroplanes. A.V. Roe & Co. Brownsfield Mills, Manchester. Consulting Engineers. Work drawings made. Aeroplanes built to clients' own designs. Monoplane, Biplane or Triplane. 5 miles flight guaranteed. Instruction In Flying. General Manager. A.V. Roe, The First To Fly In Britain.'

I couldn't believe my luck when I received a reply from H.V. Roe. H.V. manufactured surgical webbing and braces at the address I'd written to, Brownsfield Mills, and was making triplanes for his brother, A.V. Roe. He suggested that I might learn to fly in return for my services as a mechanic to his brother who was now resident at Brooklands. Would this be agreeable? *Would it not!*

What more could I have wished than to join the first Englishman to design, build and fly his own plane, who was known as *Roe the Hopper*, and to have the opportunity of going to Brooklands where the majority of British constructors had collected with their flying machines? We agreed that I should pay a deposit of £30 out of which costs for any damage I did to the machine while learning to fly would be deducted. I sent off the money. H.V. Roe sent me a receipt, 'Received 24 June 1910, deposit Thirty Pounds. To be returned when lessons complete, less cost of repairs to aeroplane.' With this settled, I sold my beloved Minerva motorbike, packed my few personal belongings, bade my dear landlady farewell – she was sorry to see me go – and with £50 in the bank and the receipt in my pocket, I was ready to leave.

What luck, but I was entering into an occupation which was considered dangerous and those connected with it looked upon as being a bit peculiar. Mr Magnier, however, was delighted to hear the good news that my longing to fly was at last being realized and cordially wished me well, thanking me for my year's service with him.

I was off to Brooklands!

1903 – 1910
Notable Petrol Powered Flights

Wright Brothers	America	1903	First to fly.
Santos-Dumont	France	1906	Second to fly.
Blériot	France	1909	Crossed English Channel.
Roe	Britain	1908	Possible hops on British plane.
Cody	Britain	1908	Sustained flight on British plane.
Grahame-White	Britain	1910	Attempted London to Manchester flight.

CHAPTER FOUR

With Roe at Brooklands

It was a pleasant day in June 1910 as I travelled by rail from Manchester to London, then 15 miles west from Waterloo Station to Brooklands. As we pulled in at Weybridge station I could see the famous motor track in front of us. It certainly looked splendid and I never felt happier. I jumped from the train and made my way towards the track, but as I came closer my impressions of the much publicised flying and the number of people who flew there was not what I'd imagined. I had the idea now that flying had arrived everyone would be wanting to learn and so expected to see great activity and Brooklands packed with machines. There was not a single aeroplane in sight.

A little dismayed, I passed under a tunnel over which part of the oval track ran, came out at the paddock and looked around. The ground lay between Weybridge and Byfleet, to the north ran the railway which had brought me here, and the track itself seemed to be about three miles around with parts very steeply banked, then within the course was the area for flying looking rather bumpy in places. The River Wey flowed diagonally across. Everywhere seemed so silent. In the distance about a mile away were the flying sheds and I set off in a beeline towards them, crossing over the little footbridge as I came to the river. After a few minutes of brisk walking, I neared the sheds and counted well over a dozen of them, and a few unfinished ones, then I spotted the one I wanted. 'A.V. Roe'. Within minutes I was there. Inside were two tri-planes, stepping further inside I found a man several years my senior, in his thirties I thought. This must be A.V. Roe!

He looked up blankly, so I quickly explained that I had been taken on as a working pupil by H.V. Roe in Manchester. It *was* A.V. Roe but his expression turned to one of surprise, even annoyance. He wanted to know why on earth have they sent me down here, he had no machine available to teach me on, and my spirits dropped to zero. After all my hopes to be greeted like this! Weren't those two triplanes suitable? I suppose he saw the expression on my face, he must have, and for a moment he looked at me searchingly and then said with more encouragement in his tone, 'Well, if they've taken you on in Manchester, I suppose you'll have to stay.'

With restored spirits I went into Byfleet and soon found moderate accommodation close to the track at the Coffee Tavern. Within half an hour I was back at Brooklands and later in the day helped Roe wheel one of his triplanes out. It was powered by a 35hp engine which we got running, but I discovered that this was quite a serious problem as the only way to warm it up was on full throttle and to prevent the machine running away we tied it to a post. Roe climbed in and when he'd made sure everything was in order, I untied the tail and the machine started to move. Enthralled, I watched from the shed door. It bumped and charged a short distance, and then I saw a gap between the wheels and earth. Was it really flying? *It was!* This was the first time I'd ever seen a machine in the air. I saw a Voisin on display in Germany, also Grahame-White's Farman at Lichfield, that's all, never one actually flying. He did a straight flight lasting a few minutes, landed and bumped his way back. I went to meet him to wheel the plane over the last few yards and manoeuvre it into the shed, then I dropped a hint. 'I should like to have a shot at getting it into the air.' Nothing doing! He ignored the hint completely as if I hadn't spoken.

I slept soundly that night and early next morning set off to do a day's work. As I entered the aeroplane shed a man dressed in a long working jacket was lying on the earth floor engaged in tightening one of the wheels of the triplane that Roe had flown the previous day. He looked up. 'You must be Pixton.' He got up and came towards me. He was Platt, a man of medium build, and like Roe my senior in years. I followed him to the other side of the machine as he continued checking the wheels. I commented that it was very quiet here, but he said, 'Wait until there're some crashes, there's always something going wrong with the machines,' and he told me that recently Gilmour had a nasty fall in his, going head over heels when a wheel caught in the telegraph wires near the paddock, and another crumbled up when the wing touched the ground and broke to smithereens. One of the main difficulties the chaps have, he said, is landing too heavily, chassis break or buckle, wheels get squashed, tyres burst, propellers split and shatter, then ailerons lock, wires snap, struts break, and whole structures twist, and sometimes the engine chooses to stop midair. Flying sounded a hazardous occupation, and from what Platt was saying anything could happen.

Platt didn't fly, he did odd jobs. He and Roe had recently moved a seat forward for better balance and had shortened lower wings and added to others. They did repairs for other experimenters too and a local carpenter helped whenever they wanted him and made up a few propellers. Platt had been in Manchester working for his brother, H.V. Roe making braces, then a section was cleared when A.V. started on his planes and a few men were taken off the lines to build them. One of Roe's planes had been named after the braces they made, called *Bullseye* and he also had a small catalogue advertising spare parts for experimenters called *The Aviators' Storehouse* all helping to raise money to support Roe's flying. When Roe came to Brooklands, Platt came too. The real danger of flying was wind, even breezes could be dangerous, one never knew when a gust would hit them, but most flying was done in the early morning before the sun got hot, or in the evening as it was more likely to be calm.

Most of the experimenters at Brooklands were balloonists or connected with the motor business. Their sheds, all very primitive with canvas doors and soil floors, were numbered and called *hangars* rather than sheds now, a French word, which was catching on. Platt took me around beginning with (1) Lane's monoplanes, (2) Voisin Box Kite which Captain Maitland sometimes took out, (3) Lane monoplane belonging to Astley, (4) and (5) the Humber Company, who had one plane which Captain Lovelace and George Barnes, the well-known motorcyclist, were trying to fly, (6) Neale's, trying to fly a machine he'd just made without much luck, (7) Gilmour's, Morison helps him, (8) didn't know much about it, (9) Alan Boyle with a Howard Wright Avis, thought he'd left now. We'd come to the end of a line of hangars, (10) taken by a dentist called Humphreys, with a machine nicknamed *The Elephant*, his weekend cottage was just behind it, known as *The Birdcage*, (11) the Petre brothers, very well-liked, the serious one nicknamed Petre the Monk, the other, Petre the Painter, always cheerful and smiling. Inside their hangar was a very clever, unfinished monoplane with its propeller at the back where the rudder should be, and a long shaft connected it to the engine at the front, it had been shown recently at the Olympia Aero Show in its skeleton form and was beautifully made. Next was (12) Martin and Handasyde, George Handasyde was better known as Handy, inside was a handsome machine with a long body looking very much like an Antoinette, no number 13, (14) was one of the larger ones and was ours, (15) Collyer monoplane, something like a Blériot, (16) Grahame-White's machines, the Farman he used on the London to Manchester flight, also his Blériot, (17) Bristol's hangar, a branch of the tram firm, their full name, British & Colonial Aeroplane Company (later Bristol Aeroplane Company), Maitland was trying out their first Box Kite. We came to the last hangar, it housed the *Flying Bedstead* of Brooklands.

The Automobile Club intended to build more sheds. Some residents would no doubt move into bigger and better hangars as they expanded. I suppose some people would leave too, for rent was high at £100 a year, more than an average man's wage, but I learnt that many paid £10 a month and shared the costs with friends. Some trees and shrubs enhanced the scenery, and a patch of gooseberry bushes flourished nearby, but the grass was a bit thin in places and the ground uneven, and then I noticed what appeared to be a small ploughed field ahead of us, its furrows wet with water – the sewage farm. Platt said there was no smell, the track was built around it before anyone imagined any flying would be done here. We returned to our hangar, there was a primus there to make tea and a couple of boxes to sit on.

A few days later, I experienced my first race day at Brooklands. Race days always brought out the planes. Several worked unceasingly into the night to get their machines in flying order on time for it was a great incentive to fly in front of crowds. *Brooklands came alive.* Hundreds of people arrived by car and train paying their entry fee of one shilling upwards, then the motor racing started. As the cars sped around, I found myself watching one man in particular who stood out by his very skill. Fascinated by the speed at which he shot around the course, I asked Gilmour who was standing near, 'Who's that in the Napier?' It was Edge who did a lot of racing here, one of the best. He must have been going over 60 miles per hour, and to think, not so long ago a scientist said '*no man would be able to drive at sixty miles an hour and breathe*'. While the racing was on, a few machines were in the air but most of the flying would be done afterwards. Some of the men just wheeled their machines over the ground to show them off to the public who were free to stand around the hangars. Interested spectators included men, women and children.

During this period a flight with Grahame-White on his Farman was auctioned. The bidding had continued briskly until most were forced to drop out and it became a contest between two ladies. The stakes rose to an incredible height, and continued to rise until, 100 guineas, 115, 120, that was it. Lady Adby had gained the privilege of flying with Grahame-White for 120 guineas, I could hardly believe it, a joy ride for 120 guineas! No wonder an excited crowd had watched every movement as Lady Adby was escorted to his Farman and no wonder a cheer went up as she was helped onto it. In common with most ladies of the day when travelling, she wore a veil and a white scarf over her head, tied under the chin and she proudly sat aloft as Grahame-White took his seat. A wealthy lady with money to spend and an adventurous one, the crowd loved her for it. The engine had started in an uncertain manner as the Farman ran along the ground for a few yards then left the ground but fouled some bushes with the tip of his wing and it looked as though he and his lady passenger had ended up in the River Wey. Everyone rushed across but it had just missed the water and was in a very sorry state on the bank. Neither Lady Adby nor Grahame-White was hurt. What happened to the 120 guineas I never discovered, but I had thought, if an experienced flyer like Grahame-White got into difficulties like this, whatever would happen to me when I'm flying? I was soon to find out.

Roe loved his triplanes. At first I wondered whether he ever went to his digs in Byfleet as I would leave him at the end of the day and always found he'd arrived before me in the morning. It was no surprise when Platt mentioned that he used to sleep in a hammock in the hangar when he first came to Brooklands. He was a persistent worker, constantly thinking out ideas for new machines and trying them out when he could on the machines we maintained. He flew a lot too, my work consisted principally in minor repairs after his flights, slightly altering parts, dismantling, rebuilding engines, all for better results. We used 35-horsepower Greens made by the Green Engine Company, one of the first British companies to attempt building aero engines. The hardest job of all was getting one to start. I had to push petrol-soaked rags into the air intakes, and keep turning the propeller until it decided to run. Whenever we wanted to test an engine's pulling power, I went over to the Bristol hangar and borrowed a spring balance, quite a heavy thing made of brass which I attached to the tail skid with strong cord, then tied the other end to our post. With this done, I would get the engine going and thus see what pounds it was capable of pulling.

I'd been working in Roe's shed for several days now and was wondering when I would get a chance to fly. I came here to learn to fly. Perhaps Roe thought I might crash one of his beloved planes? No one had flown either of the triplanes except Roe himself. I watched him every time he flew and each time I started a machine for him I could not help wishing that he would take me up for a little instruction, or at least a flight. He must have read my thoughts, for without warning he asked whether I wanted to come up. I looked at him. *My dream was at last being realized!* Without a moment's hesitation I downed tools, we got the engine going and I jumped in squeezing into position. A.V. got in and we arranged our caps in the fashionable way for flying, that is, with the peaks worn backwards, and there was a roar, the engine raced, the propeller turned, Platt untied us, *this is it!* we started to move...

I'd seen this machine bump along the ground and was surprised to find it ran extremely smoothly, the suspension so adequate that I had no idea the precise moment we left the ground for suddenly we were in the air! I felt no fear. I looked down. We were about 5ft high, and higher we went to 10ft until we were flying on an even keel at about 25ft. I never imagined it would be like this. Oil blew into our faces from the exhaust making us look like sweeps, but that was quite the normal thing. I could see for miles around, the track below us, St George's Wood to the right, the railway, the paddock... *I'd never felt such a sense of security*! But all too soon it was over. We landed, and as I hopped out I thanked Roe, inadequately, I thought.

Not long after my exhilarating first flight with Roe, I repeatedly expressed that I would like a little instruction as the agreement had been that I should give my services as a mechanic in exchange for flying instruction. A.V. relented a bit, and told me to practise rolling the older of the triplanes along the ground to get the feel of the controls, but on no account try to fly. A few people were watching as I wheeled the machine from the shed and started the engine. I sat at the controls knowing full well I was going to flagrantly disobey A.V's strict instructions. Once untied, I opened the engine full out and darted off down the field feeling very much at home at the controls. I accelerated and climbed to about 50ft, smiling grimly at the thought of what A.V. would say when I got back on the ground, but the smile soon vanished when the wings dropped on one side. Alarmed, I pulled frantically at the control column, but they dropped further, so I pulled the other way, and up they came. I'd been pulling the wrong way. After this shock I decided to land while the going was good and came to earth with a heavy thud. I turned the machine around and slowly returned to base and when I came face to face with A.V. he didn't say a thing, he didn't look pleased. No congrats from him. He hadn't approved of my flight in the least, being much too indignant about the risks I'd taken with his triplane, yet I could tell he was really quite impressed. The machine had not suffered in my hands, but I'd learnt that flying practice was harder than flying theory, that the controls needed much more handling than I'd ever expected. However, *my flying career had started!*

I was extremely lucky to be at Brooklands. It was the first centre in Britain where men in oily overalls and leather jackets collected together to work on building aeroplanes with the sole purpose of flying them. Everybody was happy, everyone friends, each bound together by the same enthusiasm, the love of aeroplanes, each with dirt-ingrained hands. But we were a mixed crowd. Some were well-off, some as poor as I was but it made no difference. There were no class distinctions at Brooklands. Some came for the odd day or two each week, others spent all their time with their aeroplanes, many having given up well-paid positions to do so. Some economised drastically in order that an engine could be afforded, eating cheaply, living in the cheapest lodgings, working long hours sacrificing everything, while others had more money than they could ever spend, but I'd never encountered such wonderful spirit or such a happy band of enthusiasts. Tools were borrowed, large and costly items, such as propellers and engines were lent, exchanged or given away as one kind was often found to work better in another type of machine. We wandered into each other's hangars for a chat or just to see what the other fellow was doing and little groups always collected when something was going on, but an exciting moment was when a machine came out for a flight.

Apart from the engine, the main ingredients for a plane were wire, wood and cloth as it was thought nothing solid and heavy could fly, but although the machines looked frail they were, on the whole, well-built and strong enough if treated with moderate care and consideration. Average speeds ranged between 30 and 40 miles per hour. One aeroplane had been built entirely from old packing cases, a bundle of calico and an old motorcycle engine, and another thrifty experimenter had used strong brown paper on the wings. Likewise, instead of varnish, boiled sago was often used as dope to make the fabric tight. It did the job pretty well, too. Willing helpers would lie on the ground to hold onto the wheels and others clung onto the tails to prevent the planes bounding forward as propellers were swung and engines warmed up. Roe's idea of having a post to tie the triplane to served us well. People came from all directions to look at the failures and successes and often it took months to get a machine to fly, if at all... many gave up and it was sad to see them go.

Within a few months Brooklands had become the nerve centre of British aviation, and here I was in the heart of it all. I cannot estimate accurately how many machines were in Britain at this time, but there were well over twenty at Brooklands though many of them could not fly, even fewer were at Shellbeach, or latterly at Eastchurch, the Aero Club's new grounds. There was nothing to compare with Brooklands in character and size but we were still looked upon as aerial lunatics, and all those concerned with aviation stood apart from the life of the ordinary man in the street. Fortunately we kept to ourselves and our way of life did not interfere with theirs, but slowly more and more people were becoming interested in our work and came to visit us with the hope of witnessing some flying.

I'd not been long at Brooklands when the Blue Bird Café came into existence. It was Eardley Billing's idea. He was a newcomer too, and had got permission to convert an empty hangar into a refreshment room, the No 8 shed which had housed a machine that had done nothing in the way of flying, and he brought his wife to run it. They nailed up a counter of sorts, filled the place with tables and chairs, and fixed a board outside with a bluebird on it. Here we assembled to drink tea, discuss flying, talk about the latest crash and new ideas, and gradually fewer primus stoves and old buckets holding coke fires were found in the hangars. Without question, the Blue Bird Café was the highlight of Brooklands. No one could have come up with a better idea. It was in a good position too, overlooking the flying area so that when we were having a cup of tea we did not miss much. Mrs Billing rather mothered everyone, bandaging cuts, sewing on buttons, and even darning socks for some. Billing meanwhile busied himself with a Lane glider, saying that anyone who wanted to learn to fly should first learn how to glide and then graduate to the aeroplane. I didn't go for this idea much, too remote, nor did I think much of an earthbound contraption he had which was supposed to simulate the rudimentary movements of flight and was even further from the real thing.

As more hangars were built, new men and machines arrived, and a recent newcomer was Lieutenant Gibbs who had learnt to fly a Farman in France and had just received his English licence, No 10. He came to Brooklands with some strange stories to tell of his experiences in Spain, one of the many countries which had not yet started in aviation. He'd signed a contract to fly his Farman at Durango with a French pilot flying a Blériot, but the Spaniards burnt both planes as they thought they were the work of the devil, he was hardly believed... but it turned out to be quite true. He told the story:

> 'We assembled the machines but the crowd started to throw stones at us so we shut ourselves in the sheds, but they cut holes through the walls to get at us, and a group was going to break down the doors with a ramming pole. I was scared stiff when a man pulled out a knife and told me in bad French that he was going to kill me and that there was no such thing as flying. The crowd were chanting "*down with science, long live religion*". Eventually a mounted guard at the ground charged them with drawn swords and we left under their protection on being assured that our machines would come to no harm. Some assurance! We found the sheds burnt to the ground with our machines in them. Nothing was saved.'

It all sounded very primitive, but not long after Gibbs's amazing story we heard that a crowd at Worcester had tried to burn a pilot's machine because he did not fly. Probably the weather was not right, then at Halifax it was reported that Grahame-White had had bricks thrown at the shed containing his machine when he refused to fly in a dangerous wind, and at Crystal Palace a well-dressed crowd booed someone else for not flying. Nothing like this had ever happened at Brooklands, thank goodness. Another newcomer to Brooklands was Macfie who came into a shed behind us. He was engaged in building a machine with very inefficient equipment. He built his first machine at Fambridge, then went to Foulness Island to fly it but War Office officials chased him away so he went to France to set up there, couldn't find a suitable place so returned to England and had discovered Brooklands.

I grew increasingly happy and contented as the days went by, work was a pleasure, time passed pleasantly, I was out in the country and lived nearby. I always found it very pleasant wandering around the ground. The River Wey running across Brooklands where punts were often moored, added to the atmosphere, and when all was quiet I would see rabbits darting about and vanishing down their burrows. However, it was a long walk to the aeroplane sheds for the men coming in from Weybridge, and this particular stretch from the motor paddock was being called by them '*the desert march*'. Gilmour owned a dog, an intelligent-looking animal called Seti which was half wolf. He found him wandering about in the Tomb of Seti in Egypt's Valley of the Kings practically starved and brought him back to England and kept him in his shed usually on a chain, but if anyone went near his shed he became very excited. I'm afraid I teased him every time I passed by putting my foot around the door. That really upset him, invading his territory, but I paid for it. One day I suddenly felt a sharp nip in the leg and promptly turned and before I could do anything about it, poor old Seti was running as fast as his legs would carry him, his deed done, having picked me out and got his own back.

Everyone was shocked by the very sad news of the first air fatality in a powered aeroplane to occur in Britain. It happened on 11 July 1910, at the Bournemouth Air Meeting. The Honourable C.S. Rolls, well-known as a pilot, had died in a crash. Cody, a close friend of Rolls was there, so were Grahame-White and many from Brooklands – Gilmour, Gibbs, Boyle, Astley and Radley. We from the Avro shed were not as Roe was preparing for the forthcoming flying meeting at Blackpool. Radley and Gilmour gave us an account of the accident on returning to Brooklands. Competitors were trying to land as near as possible to the centre of a circle, Grahame-White landed near the edge, Dickson landed outside it, both had a side wind, but Rolls tried it over the grandstand so that he would land into the wind. He realised the angle was too steep and sharply pulled up. His tail broke off. He somersaulted over with the plane on top of him, was doubled up, his neck broken, and as the crowd rushed into the flying area many were weeping. Gilmour told us that no more flying was done on that day and that the rest of the meeting was an ordeal, having to fly as though nothing had happened.

Rolls was a founder member of the Royal Automobile Club and the Aero Club. At the turn of the century he had joined forces with Royce, who'd gained recognition in the production of the Royce motorcar renowned for its silent running, and the firm became Rolls Royce. Although Royce was not very interested in flying, Rolls pursued it as a pastime with great enthusiasm. Much of his flying was done at Eastchurch. He was considered to be one of the best pilots in this country and a most unlikely person to be killed in an aeroplane. Only a few weeks before he flew from Dover to Sangatte and back without a landing, so becoming the first man to fly the double Channel crossing. Aviation had lost a fine man. Rolls's death occupied the thoughts of thousands and caused a great public sensation. Hundreds came to his funeral at Llangattock, Wales. Then there was the outcry... *'Flying must be stopped.'* No such step was taken. Although several French pilots had been killed flying, Rolls's death brought home to us that flying was a dangerous occupation and that anyone could be killed by it. Moore-Brabazon, most upset over Rolls's death, virtually gave up flying, complaining about the entertainment side of aviation at flying meetings. He and Rolls were very close friends.

I had been the first man to fly one of Roe's aeroplanes apart from himself. He might have taken up other people, but I don't think anyone had flown solo on one of them. Since my first solo, I'd been doing a few straights then a young man came to learn and was accepted by Roe who had in mind to start a flying school. The young man was with us only a couple of days and was not seen again after damaging one of his machines. As the result of my flying a bit, my name began to appear in the flight magazines. On 13 July 1910 it was reported:

'Mr Pixton, a pupil of Mr Roe's, took out the passenger machine making some short hops and bringing the machine to ground in a way that promises well for his future progress, though on his last attempt he came down rather heavily from a height of 15ft and slightly damaged the wheelbase.'

It was a great moment when A.V. came up with me as a passenger, for this was the only time he'd ever allowed himself to be piloted by anyone. This was also reported.

'Mr Roe was out as usual and had the novel experience of travelling as passenger on one of his own machines, Mr Pixton being at the helm. Mr Pixton is rapidly getting control of the machine, making steady straight flights and bringing her to earth as gently as his tutor.'

It was then I suspected that A.V. was not over enthusiastic about flying, that his real interest lay in design and invention, and I thought I might become quite an asset to him as a pilot in later months when I would have gained more flying experience, thus having the opportunity of repaying him. He'd opened the doors of a flying career to me, something I could never forget.

During July A.V. was watched by Aero Club officials as he completed the last of the flights for an Air Licence. No one was really in a hurry to secure a licence since we could fly without one. So far Radley, Boyle, Barnes and Roe had been the only ones to get theirs at Brooklands, and only a month previously Cody passed his tests at Laffan's Plain. Although both Roe and Cody were the first to fly British-made planes in Britain, Cody's licence was No 9 and Roe's No 18. The following are those who had received licences, the planes they flew and where, between March 1910 and July 1910.

1910
First British Air Licences

	Pilot	Plane	Place	1910
1	MOORE-BRABAZON	Short	Shellbeach	March
2	ROLLS	Short Wright	Shellbeach	March
3	RAWLINSON	Farman	Shellbeach	April
4	GRACE	Short Wright	Eastchurch	April
5	COCKBURN	Farman	Mourmelon	April
6	GRAHAME-WHITE	Blériot	Pau	April
7	OGILVIE	Short Wright	Camber	May
8	MORTIMER-SINGER	Farman	Mourmelon	May
9	**CODY**	**Cody**	**Laffan's Plain**	**June**
10	GIBBS	Farman	Mourmelon	June
11	EGERTON	Short Wright	Eastchurch	June

12	RADLEY	Blériot	Brooklands	June
13	BOYLE	Avis	Brooklands	June
14	DREXEL	Blériot	Beaulieu	June
15	COLMORE	Short	Eastchurch	June
16	BARNES	Humber	Brooklands	June
17	DAWES	Humber	Wolverhampton	July
18	**ROE**	**Roe**	**Brooklands**	**July**

    A.V. was going to attend the Blackpool Meeting at the end of July. I would go along too, both planes would be taken. This was the second flying meeting at Blackpool. Roe had attended the earlier one during 1909 so he was not new to the procedure but, for me, it was to be the first time. The triplanes were in splendid tune as we dismantled them, ready to be put onto the train. He and Platt set off to Blackpool, leaving me behind at Brooklands to lock up and attend to things generally. The following day the railway company carriers came, and I went to Weybridge station to see that everything was in order as they loaded the planes on the railway truck. With them went spare parts, two bicycles and many pieces of A.V's personal belongings, including clothing. In fact, nearly the complete contents of our shed went onto the truck, and the whole lot was carefully covered with a tarpaulin sheet for their journey to Blackpool. I was quite satisfied.

    Next day I was ready to depart myself, and took a train from Weybridge to London, then caught another from Euston. I'd settled myself down to the nice, peaceful journey but halfway between Chorley and Preston the train unexpectedly pulled into a siding and stopped. Bewildered passengers leant out of the windows. I was curious to see what was going on too, and jumped down onto the lines when I saw smoke billowing from the front of the train. My curiosity turned to horror! I couldn't believe my eyes. There, at the front next to the engines, were our aeroplanes blazing merrily. Everything else was burning too, the bicycles and clothes… and all the time I'd been sitting comfortably in the train with an easy mind without even knowing the planes were on my train. I'd imagined they would have arrived at Blackpool, but apparently they'd lain overnight at the station in London and had been hitched onto the train that I happened to be travelling in. Sparks flying from the engine had set the tarpaulin alight, and the whole truck was on fire before anyone had realised. Nothing could be done. The fire had secured too strong a hold and was making ashes of the only triplanes we had. How could I tell A.V.?

    As I stood watching the truck being shunted off and left to burn itself out, I felt responsible to Roe even though the planes had not been left directly in my charge. I'd done all that should have been sufficient by putting them into the hands of the railway officials. Filled with resentment at the carelessness of the railway company in attaching them next to the engine, I climbed back into the carriage. Why on earth next to the engine? Two elderly ladies in my compartment asked anxiously, 'What happened?' I told them that some aeroplanes at the front of the train for a flying meeting in Blackpool had caught fire. One replied. *'Well that's something to be thankful for. It'll save somebody from being killed on those horrid flying machines.'* It was obvious that I could not express what a tragedy this was, so said no more. They meant well, and were only expressing the general opinion that those concerned with flying were irresponsible men out to kill themselves. There was no pleasure in the rest of the journey.

As we drew into Blackpool station the first person I saw was A.V. awaiting his machines. Apparently he'd been notified of the time of their arrival, but I had the painful duty of telling him what had happened and broke the news as briefly and as directly as possible. 'Both planes are gone, burnt. They caught fire on the train.' I've rarely seen anyone so upset. He was a complete picture of despair, and I just looked at him not knowing what to say, but knowing anything I might have said would have been of no comfort. He felt he was right back to where he started... *All lost in one day!* For a while I thought that A.V. would pack up flying for good and go back to the braces factory to help his brother, for they'd spent a lot of money on machines and it was obvious that there was a definite shortage of funds. Roe was hoping to win some of the prize money and bring in orders. He wanted a market for the triplane but now he had nothing. The planes were carried at one's own risk. There was no insurance. No insurance company would look at an aeroplane, just as they would not insure a pilot's life, the risks were considered too great. There had been another plane on the same train, a Short owned by Cecil Grace; his had gone unscathed.

This was a disaster that nearly lost to the world the distinguished name of Avro but I'm glad to say Roe recovered from the bitterness of the blow and was determined to compete at the Blackpool Meeting. The meeting started the next day but without hesitation we travelled on the next train to Manchester leaving Platt to answer for our absence. Once in Manchester money was raised through the faith of his brother and his manager, John Lord, and we went to work without a moment's delay. Production of braces and webbing came to a standstill as all hands helped. Out of spare parts we started building a new machine, but not without difficulty in finding some parts. We could not get an engine to Manchester on time and had to arrange for one to be sent direct from the makers to Blackpool; the tyres did not fit onto the rims of the wheels satisfactorily, but they had to do; the fuselage was left uncovered, we had no time. After working for three days and nights, we had a complete spare-parts triplane, and we rushed it to the station, making certain it was not put behind the engine, then we boarded the train ourselves, all set for Blackpool. Building a plane in three days and three nights, put me in mind of the joke about constructors, *'Yesterday I designed her, today I built her, tomorrow I fly her.'*

We arrived late on Sunday, halfway through the first week of the meeting. Platt said we had not missed much. Thursday, the opening day, was a flop. A few machines were unpacked but no one flew and the angry crowd was given free tickets for the next day. Grahame-White flew for about twenty minutes late evening when most of the people had left and got about £200 for that. Each day prizes of £100, £50 and a further £50 were given for the longest flights of the day and for the greatest altitude. There was also a £100 prize for the most meritorious performance, a total of £300 a day could be won. On Friday a notice had been put up saying flying would take place if possible and that no money would be refunded. Again no one flew, too windy, but a bit had been done on the Saturday and Sunday.

That night I slept in a packing case in a hangar, partly because I had no lodgings, but mainly to see that nothing should happen to the machine. Early morning someone woke me by hitting the packing case with a hammer, and he hit hard. I couldn't really complain as it certainly got me up. It was a glorious day, a Bank Holiday Monday, and thousands turned out to watch the day's flying, the best day so far. Our hangars were open to the public from 10am to 2pm, from then onwards was the official period for flights, until 8pm, but we in the Avro shed shut ourselves in since the engine had arrived and we were busy fitting it. With a certain amount of pride we announced the triplane was ready for the afternoon's flying.

When we made our appearance, the crowd became very excited. We almost heard them gasp with surprise as our loss had become known. Then they cheered. A.V. took his place at the controls and they cheered even louder. I'd never known such popularity but Roe as a Lancashire man was well-liked here in Blackpool. I could hear cries of well-wishing above the general excitement, *'Oop, Lancashire... Ah, he's a gradely lad.'* Roe took off in good form, did four circuits with some impressive sharp turns amid the cheers and clapping, then landed heavily. The tyres blew off and he broke the front skid but that was easily repaired. He'd done very well as the machine was completely untested and it had flown first go. For this flight, and the fact he'd managed to produce this machine in such a short time after having lost two planes, he received a very special merit prize of £50. To my mind, it was the most well-deserved prize of the whole meeting. H.V. Roe was also there and insisted that I took a flight in the triplane. A.V. seemed reluctant, but I flew an uneventful straight flight before returning it to him. Many others flew this day, McArdle, Chavez, Drexel, Grace, Loraine and Grahame-White who flew to Southport, 15 miles south, landed there, attracted crowds and returned to compete in several events. Loraine, who was an actor known as Jones when flying, also flew off in the same direction but caused a good deal of anxiety when no news was heard of him for some time, then a telephone message was received, *'Jones seen flying over Liverpool'*.

All this flying had no restraining effect on me, quite the reverse in fact, and the wild idea of flying around Blackpool Tower entered my mind. I spoke of this to Albert Pratt, a very old friend of mine who was, I discovered, in charge of the proceedings at Blackpool. He sympathised. It was certainly tempting, and next morning I got up early with this intention in mind and was met by Albert who was going to see it through with me. 'It's too windy,' he said and was right. Somewhat relieved I knew I had not enough flying experience, not having yet made a left or right hand turn, although they looked easy enough... it had been rather rash of me to think of it. The day turned out to be wet as well as windy, and the flying ground was practically deserted.

Roe went out in the evening and was out again the following day. On this occasion he covered practically half the circuit before lifting, and then only did two rounds, came in to land and broke the axle. I got that fixed. He went up again but was caught in a gust and crashed nose first to the ground. He was none the worse for it, but he'd never done so much damage in so short a time. And so, the first week of the Blackpool Meeting came to an end, Grahame-White claimed £650 in prizes, Chavez, who took an altitude prize and established a new British record of 5,850ft, came next with £225. Unfortunately during the second week, a Scottish Flying Meeting was on at Lanark and most of the pilots went there to compete, so there was not much doing at Blackpool. Grahame-White, engaged for a fee of £2,200, gave exhibition flights for a few days along with Tetard, a Frenchman engaged at a more moderate rate. Attendance was very low.

As the Blackpool Meeting came to a close we were rejoined by most of the competitors from Lanark and further prizes amounting to £2,000 were awarded. Grahame-White gained another £490. We had had shocking weather, apart from the Bank Holiday Monday, and the Lancashire club who'd arranged the meeting was sadly out of pocket. Land, enclosures, hangars, officials, fees and prizes all came to a great deal of money which had to be found somewhere. A lot was given privately but much depended on good attendance and good attendance we did not have. However, it had been an exciting time for me. I'd flown a few straights and met many pilots whom I'd only heard of before. The main outcome for Roe was an invitation from J.V. Martin of the Harvard University Aeronautical Society, to visit Boston and participate in their first flying meeting due to be held between 3-13 September 1910. He was given an order for a triplane, and in spite of our financial difficulties and the fact that we had only our spare-parts machine, an agreement was reached. Roe wanted me to come with him. I'd been to Germany and now America! Grahame-White was invited and he would also be taking along helpers. Bookings were made and Roe alerted his brother. *'We urgently need another triplane.'*

I'd made friends with two men, Oxley and Sims, who had jogged merrily along with us throughout the meeting as voluntary helpers only too pleased to lend a hand. Oxley, who'd come all the way from Yorkshire just to be at the meeting, was making a living selling goods from a stall and wanted to fly. He was very keen at the idea of coming to Brooklands to do so. So was Sims. I told them if they were seriously considering flying as a career, then there's no better place than Brooklands. *'Does tha think we cud coom back wi' you, lad?'* We were going to America and not returning to Brooklands until October, so I suggested that they should come about mid-October. Roe was in good spirits as, despite his losses, things were going well for him and on the final day of the meeting he married. The ceremony took place at South Shore Parish Church, Blackpool, his bride, Miss Mildred Kirk. After the wedding, we rushed to Manchester with only three days to spare. It took some time to get another machine ready but it was given essential priority by the factory and within record time it was completed. With our spare-parts old faithful overhauled and the new one packed, we were ready to leave.

We set sail on Tuesday, 23 August 1910 bound for Boston on the White Star Liner *Cymric*, once the world's largest cargo boat which became one of the earliest to be fitted with first class accommodation. A new man, Halstead, came with us as part of A.V's staff and we met up with Grahame-White, Sidney MacDonald and their three men, Carr, Turner and Hucks, the wealthy man of the party. A.V. Roe, Grahame-White and MacDonald travelled first class, the unfortunate mechanics steerage. To begin with, Grahame-White's men presented a different atmosphere to the one I'd been accustomed to at Brooklands. One might say they were such *haughty mechanics* who seemed to be trying to get through to Halstead and me that we were privileged to be in their company. Posh? There were no privileges in travelling steerage on this trip and by mutual suffering we quickly became friends.

Steerage in 1910 on any ship was bad, very bad, and the journey turned out to be quite an ordeal. I'd never experienced anything like it. Never have I been so utterly miserable! It was Hell. First of all we were not permitted to get on the boat at the Liverpool landing stage but had to go out by tender. Then, once on board we were lined up for an examination by a doctor, to which I strongly objected. I didn't want to be checked for diseases and have a doctor stare down my throat and peer into my eyes, and managed to miss it. While the MO was looking down one man's throat, I slipped behind his back and joined the passengers who had already been examined.

If I'd known what I was in for, I would never have come. The berths were hard, meals were far from good and the manners of many who ate them were worse, so every day I slipped off, unobserved, into the first class and joined A.V. My luck ran out one day. About halfway across, a stern-looking officer in a peak cap caught me as I slipped through into the first class. It had to stop. As we neared Boston, the MO had his bit to do again, this time the vaccination of steerage passengers. I took a very dim view of this and tried to conjure up a means of dodging it, but evading it seemed out of the question until Carr hit upon the idea that the five of us could go into the hold on the pretence of overhauling an engine or something. It worked. We stayed below with the planes, oily rags in our hands and tools laid out in case anyone bothered us, but no one did, then Hucks slipped up on deck and came back with the news, *'Operation Vaccination's over'*.

Boston came into view and as the *Cymric* went into dock, a motor launch belonging to Mr Martin met us. It drew alongside and a rope ladder was thrown down to it 40ft below and we were expected to climb down on it. Once aboard we were taken down water to Boston docks but our troubles were not over. On arriving at the Squantum Flying Ground a few miles out of Boston, we discovered that the only sleeping quarters for mechanics were canvas hangars. I selected my spot under one of the triplane's wings, but each night we were invaded by what seemed to be everything that could fly, crawl or sting. There was also a continuous chorus from crickets as we tried to sleep. Next day we discovered we were near a bog and it was from there that these creatures came. As we wandered around, taking in our bearings, one of Grahame-White's mechanics disappeared chasing something, a beautifully marked cat with a bushy tail. It was a skunk.

We discovered American aviation was very backward. Apart from a few firms making machines, such as Wright, Curtiss and Burgess, there was little done and the aviation meeting gave thousands of Americans their first chance to see an aeroplane. The visit of our English airmen was a great attraction. So great was the excitement and interest that people talked of nothing else and when it opened most of Boston came to watch. The university had prepared special competitors' medals and I managed to get one which I pinned onto my jacket, although I was not actually competing. They were quite large and most attractive with a Wright biplane overhead in a rural setting, and read, 'Competitor. Harvard & Boston Aero Meeting, 3 to 13 September 1910'. Quite a considerable stir was caused by the appearance of the triplanes for no one in America had ever seen anything like them. But misfortune followed Roe. Before a big, expectant crowd whose enthusiasm rose to fever pitch, A.V. took off in one of the triplanes, climbed to 60ft, stalled on a turn and made a rapid descent to earth right in front of the grandstand. We rushed over to him. This time there was nothing light-hearted about the crash. It was the worst of his career, one in which he came as near as anything to being killed. We found him bleeding very freely from cuts to the head and he was rushed to hospital, but the crowd feared the worst and the *Boston Journal* wrote of his accident:

'Roe's machine was seen to career crazily. It turned more and more on its side and then crashed to earth. The grandstand seethed with crazed humanity. There was a start to cross the field but the Police held the crowd in check. That aviator Roe might be alive never occurred to the crowd, and it was a long twenty minutes until word came that he was not seriously injured.'

Well of course he was seriously injured and the only thing that had saved him from death was having the engine in front. After he was taken to the hospital we had the job of clearing the wreckage off the flying ground. I've never seen so many people so keen to secure souvenirs. There was a rush and we practically had to fight them off to get the wreckage back to the hangar, or else we would have had just the skeleton left in our hands, if anything. Even then it was not safe. One huge man came up to me and asked, *'Can I have that little bit with the blood on it'*, and he found himself outside before he had time to realise how much bigger he was than me.

The hospital could not contain A.V. for long, and after sending his young wife a cable assuring her of his safety, he was back at the meeting determined to show the Americans that his machines were good, and took out our second plane – and crashed it! – without injury to himself, so Roe became the object of much admiration and sympathy of the Boston public. We didn't do much more at the meeting. However, that machine had been intended for the Harvard Aero Society and all we had now were two broken ones.

I saw Wilbur, one of the Wright brothers, but did not see him fly as he employed two pilots, Brookins and Johnson, who were most daring and very skilled. I thought a crash was imminent when I saw Brookins turn with a severe bank. But no, he turned beautifully. This was the first time I'd ever seen an aeroplane bank, flat turns were the rule, and banking had never been attempted either in France or England. He did it several times, and I stored away this convincing development in flying for future reference.

We had a distinguished visitor at the meeting, President Taft, one of the largest men I'd ever seen, 23 stones. A.V. and Grahame-White had long talks with him. None of the American planes could compete with Grahame-White on his Farman and Blériot machines, and he was mobbed by autograph hunters every time he appeared. He emerged the most successful flyer long before the meeting ended, but a generous offer was made to allow the Americans the opportunity of regaining some of their lost prestige by a wealthy member of the Harvard Aero Club who put up $3,000, about £600, as a prize for a race between Glenn Curtiss on his biplane and Grahame-White in the Blériot. Both men were keen to compete against one another. Mechanics and pilots made bets, but I was not among them as my pockets held but a modest amount, though I awaited the big event with just as much interest. The event was over three laps covering a total of 5¼ miles. We watched both men do their circuits individually, then the voice of the announcer came over the loud speaker, 'Grahame-White's record for the three laps, five minutes and forty-seven seconds. Glenn Curtiss's record, six minutes four seconds.' Our man had won by the breathtaking difference of seventeen seconds, and as Grahame-White appeared, the wonderful American crowd cheered him repeatedly. He certainly knew how to fly well. Prizes totalled more than £9,000 and out of this Grahame-White's triumphs had netted him a total of about £6,420, which included £2,000 from the *Boston Globe*. Unfortunately the machines he used were not British. He was to stay on in the States, joined by Ogilvie and Radley, to represent Britain in the Gordon Bennett Aeroplane contest, the second of its kind, the first having been at Reims in 1909 when Glenn Curtiss won the trophy. This year it was being held at Belmont Park, New York, in October after the Harvard Flying Meeting.

I asked Grahame-White had he heard anything more about the Wright brothers defending their patents on flight control. They had taken action not only against Curtiss, but against anyone flying a machine in America which infringed their patents. Although courts seem favourably inclined towards them, it was announced they would take no action for machines brought into America to take part in the Gordon Bennett Race, but insisted on a nominal fee to uphold their rights. Their licensees in France had recently taken legal proceedings against Blériot, Farman, Esnault-Pelterie, Santos-Dumont and the Antoinette people. I didn't know the outcome of it, but the tables had been turned somewhat on the brothers not long ago in America when Charles Lamson commenced a suit to restrain them from making machines which infringed patents taken out on his kites in 1901 relating to the warping on the wings. It seemed a legal 'mess up', and as far as I knew neither Roe nor Grahame-White paid anything to them during this visit. No wonder America was backward in aviation.

The meeting ended. Halstead was not going to return home but was hoping to make his way in America. I wanted to get home but before I could go Roe gave me a job to do, to build a sound machine from the two broken ones and present it to the Harvard University since we promised them a machine, then sell the spare parts. So Roe left for England without me. Halstead and I got busy and soon a new machine was built. It was quite easy to do as machines were not at all complicated in their construction, then we tested the engine, the better of the two, found it would run, and handed the 'new' triplane over to the university. They were very pleased to have it. Halstead then left me and I was all alone to sell the spare parts, and as I had no money, it was a dire necessity to sell them in order to survive and, indeed, to pay my return passage. Much now depended on my untried qualities as a salesman as to whether I got back or not. I need not elaborate on how energetic I became. For want of cash, I stayed on at the flying field entertaining the mosquitoes each night. Luckily, I got used to the crickets filling the air with a continuous chirping singsong, and even found they soothed me to sleep. At last I sold everything and went to New York by train, and crossed the Atlantic in the *Carmania*, this time second class and in comfort. Had I been faced with steerage, I think I would have gone back to the mosquitoes and crickets. I arrived at Brooklands at the end of October, glad to be home.

During my absence there'd been some notable achievements and a few accidents. Mosiant, a Chicago-born American of French descent, had set out on a flight from Paris to London with a passenger, but the journey had taken three weeks. Even so, it was rated as the first flight between the two capitals and was the first Channel crossing with a passenger. Wind, rain, and mechanical troubles prevented him from making much progress but finally he reached Crystal Palace. The plane used was a Blériot, the best machine of the day.

Shortly afterwards Loraine, the actor pilot whom I'd met at the Blackpool Meeting, achieved a heartfelt ambition by crossing the Irish Sea. He'd spoken about this to me, and I was very pleased to hear he'd succeeded in doing what he wanted to do. Loraine set off in his Farman from Holyhead and headed due west to Dublin, but during the flight he had some alarming experiences. The bracing wires of the Farman started snapping all around him, fortunately there were plenty on a Farman, then he was bothered by his engine cutting out and restarting, and it failed altogether not far from the coast. Man and machine sank into the water and he swam the rest of the way to Baily Lighthouse. Although he didn't quite make it all the way by aeroplane, Loraine was nonetheless credited with being the first man to fly the Irish Channel.

While I was actually travelling from America, Grahame-White took the Gordon Bennett Trophy at Belmont Park. He also won a contest from Belmont Park to the Statue of Liberty and back, and created a great sensation by landing his Farman in Executive Avenue, Washington, outside the White House. Moisant, who was not long afterwards killed in an accident, had returned to America after his Paris-London flight, and came second to Grahame-White in the Gordon Bennett contest, and had also competed against him in the race around the Statue of Liberty. And sadly Chavez, the Peruvian we'd met at the Blackpool Meeting who did so well there, had died while attempting the first crossing over the Alps. He started from Brique and was fatally injured on his descent at Domo d'Ossola, 35 miles from his starting point. There'd been no deaths in Britain, only a few broken bones. Maitland crashed his Voisin on Salisbury Plain, had had two toes amputated and was later involved in another flying accident breaking both his ankles, Rawlinson received a compound fracture of the leg which did not heal and had to be rebroken twice, and George Barnes was suffering from a fractured skull and broken arm, but there'd been a most terrible flying accident involving an Englishman in Italy, the first mid-air collision.

It occurred at the beginning of October when Captain Dickson, whom I'd also met at Blackpool, was very badly injured at a flying meeting in Milan. He was one of the earliest pilots, having learnt to fly in France at the Farman School, Mourmelon-le-Grand. Ironically, he had been most concerned about the probability of flying accidents at flying meetings, and his fears became evident in an accident in which he himself was involved. He had become very well-known for his glides from high altitudes and during the Milan Meeting was taking off on his Farman as a Frenchman, Thomas, came in to land. Thomas, in an Antoinette, crashed into Dickson's plane. Neither man heard the other because of the noise of the motors, neither could see the other as visibility from the Farman is practically nil upwards as it is downwards on the Antoinette, and both men fell to the ground with their machines locked together in a tangled mess. Dickson was found unconscious and suffering from grave internal injuries, a broken leg and dislocated pelvis. He was expected to die, so serious was his condition. And he was blamed for the collision. It was reasoned that since he was making an exhibition flight, he was more at fault as Thomas had been taking part in a competition. Thomas, unhurt, was claiming damages as Dickson lingered on for weeks, clinging to life. It was a miracle he lived, but slowly he showed signs of recovering from his dreadful injuries. His sister, Mrs Will Gordon, went to Milan with English nurses to attend to him until he was fit enough to be removed and when at last Mrs Gordon brought him home, he was still a very sick man.

It was good to be back, I returned with the Boston papers carrying reports of the Harvard Flying Meeting and must have been the one to introduce *stunt* to Britain, a word unheard of in this country, which began to appear frequently in reports. We often joked about journalese, especially that ridiculous term applied to flyers, *Intrepid Birdmen*, but *stunt* was different. Soon it was in universal use at Brooklands with the pilots and the racing car drivers, for displays of skill. I'd been away for quite some time, and much had changed at Brooklands. Many unfinished sheds were completed bringing them up to a total of thirty housing about fifty machines, but Major Lindsay Lloyd had decided that no further ones would be built as they took up valuable flying space, he did not want Brooklands overcrowded. Platt filled me in with the details. The old cottage had been demolished, and much work done levelling the ground, with as many as thirty horses and carts removing the bumps and filling in a loop in the River Wey so that it almost flowed straight across the ground. Nothing had been done with the sewage farm. The Corporation wanted thousands of pounds to remove it, the Automobile Club couldn't stretch to that type of money. The Blue Bird Café had changed too. No longer did it look like a shed with its new glass front, and inside were several improvements, a fire now winter had come and a piano was installed. Tennis courts near the clubhouse on the paddock side were planned, but that wouldn't affect us, the racing enthusiasts kept to themselves. Our helpers, Oxley and Sims from Blackpool, had also arrived and had started as mechanics with us. There were plenty of new faces too.

I soon settled back into the Brooklands way of life and got to know the new arrivals who had taken sheds. Some of the old residents had moved into larger and better ones. The new men included Spencer of the well-known balloon firm of that name, who had a biplane of his own design that he was flying from time to time. With Stirling he had designed the Spencer Stirling monoplane which they'd shown at the 1910 Aero Exhibition. There was Singer, a yachting man, a balloonist and member of the sewing machine family, but he had a plane at Eastchurch and flew there more than anywhere else, and Spottiswoode whom we'd seen at the ground several times before with friends, now learning to fly an Avis. Then there was Howard Flanders who had become so enthusiastic constructing a monoplane and a biplane that he and his helpers were living in the shed where they worked. Others included Gordon England, Valentine, Raynham, Conway Jenkins, Molesworth, Poynter, Edwards, Greswell, Watkins, Davies, Wickham, Mrs Hewlett, Monsieur Blondeau, Henri Péquet a French pilot, and many more. Péquet, in charge of the Humber sheds, seldom got his machines in the air. They were never the right type to fly, and in his broken English we would hear him sadly describe his job, *'Plenty money! No fly!'* With him hoping to fly were the Pashley brothers, Eric, and Cecil who was 19, the youngest member of Brooklands.

One the most kind-hearted men I had the pleasure to meet, was Ellis Victor Sassoon, millionaire and friend of royalty, who went under the name of E.V. Smith when flying. He had a heavy side-by-side two-seater Blériot Sociable called *The Big Bat*, and on nearly every occasion he flew, he crashed. I would see him bring the machine over the sheds, stall and flop, *The Bat* would sink onto the ground like a broody hen, and he would say, *'I'm not a very good pilot'*. At other times he completely misjudged the distance as he came in to land, usual damage, buckled wheels and broken undercarriage, but with his customary good natured grin, he would climb out look at the damage and sigh, *'I could have sworn the ground wasn't that close'*.

Tommy Sopwith, a young man of about 22, was also new. He had become interested in flying while out sailing one day when he put in at Dover and learnt that an aeroplane was in the vicinity, found the pilot, got his autograph to discover it was signed Moisant who'd flown from Paris. This led him to come to Brooklands where he paid for a passenger flight, then he bought a Howard Wright monoplane and decided to join us at Brooklands to learn to fly. He was mastering the rudiments of flight with some success, but had recently crashed the Howard Wright beyond repair and had bought another, a Howard Wright biplane this time. In his shed, Sopwith kept a bear called Pooley which was looked after by a local lad who'd just left school. Pooley was quite small, a friendly little thing, which took a delight in running up the posts and onto the rafters of his shed and there would stay for long periods on end. No form of persuasion could bring him down, until we were told of the secret to lure him, condensed milk, one of my weaknesses too, and that did the trick. With condensed milk we were Pooley's friends for life and he would wander into our hangars in search of it.

Radley, who'd arrived soon after Roe, was still with us and had made a name for himself as a daring flyer. He owned a Blériot on which he obtained his licence, an early one, No 12 and Morison was still with us, an electrical engineer and motorist who'd been flying a Blériot on which he hoped to get his air licence. The two brothers, Petre the Painter and Petre the Monk, were now busy constructing an Antoinette-type machine after abandoning their cleverly designed aeroplane with the propeller in the tail, which was a pity as it was a remarkable little machine. Grahame-White had left Brooklands. Several people were 'here today gone tomorrow', only a few were serious residents. Gilmour, a very lively person, was still around. He'd led a party of Brooklands men dressed in mechanics overalls into the Coventry Restaurant in London's Leicester Square, and caused a bit of a flutter among the smart patrons. He was not without money. After leaving college, he'd gone on a world tour, this was when he'd picked up Seti in Egypt.

There were several new Bristol men on the scene learning to fly, Low, Snowden Smith, Sydney Smith and Captain Wood. Collyer was now rolling out the *Collyer* powered by a 20-horsepower JAP. He'd renamed his creation *Ot-as-el*, rather a silly name to give to a machine, I thought. It never flew properly. Humphries, the dentist and one of the many professional men who were amateur enthusiasts, was still spending all the time he could with us. And another professional man, an artist called Fisher, had joined us and was learning to fly. Fisher had long, flowing hair, befitting only to men of his profession in this period. Martin and Handasyde, two of the earliest men to settle at Brooklands were flying their Antoinette-type monoplane. The first plane they'd built was put together in a ballroom at Hendon, a monoplane with a three-blade propeller and a Beeston engine to power it, but when they went to test it, the engine was torn out of the machine, too much power. They then designed the new version of the Antoinette, one of the first machines I'd seen on arriving at Brooklands when Platt took me around. It was the prettiest aeroplane at Brooklands, built without the usual, ungainly flaps on the wings that are seen on the French Antoinettes. The handsome spread of wing and the very long tail made it look like a magpie in flight. A unique feature about these planes was that they had wheels on either side of the pilot for balancing, one for moving the ailerons, one for the rudder. We had huge levers beside us.

The feeling of friendship grew at Brooklands. Although everyone was friendly we didn't use Christian names much. I was being called by several nicknames, Pixie, Pick, Picky Pixton or just H.P. Moore-Brabazon, though seldom seen at the ground, was to us all Brab, or No 1, since he held the first flying licence. Because of Sopwith's initials (T.O.M) and his youthful appearance, he was generally known as Tommy, and Gordon England was often called by his first name Eric. What was very noticeable about flying men, was that they were more than often the most gentle and mild mannered of men – the big, dashing fellow was conspicuous by his absence. We came to know the representative of *Flight* who visited us reporting on-the-spot news, and C.G. Grey of *The Aero* magazine, an outstanding man of Irish descent, extremely kind by nature. He did much to publicise and praise aviation, and was the best-known and the most popular journalist of the day, a pleasure to know him. Week after week he came to Brooklands, equipped with a press camera, chatting with everyone and often joining us in the Blue Bird for tea. His full name was Charles Grey Grey but he was known to one and all as C.G. Grey. I got to know him well. We often found ourselves wandering around together, talking on every subject and the news of the day, as our friendship deepened. He would say, '*When a machine looks right, it is right.*'

Roe was now living in Putney with his wife so that meant he was travelling about 25 miles each day to come to Brooklands. As he'd left Boston without an aeroplane to his name, another one was produced which had just arrived, and the engine this time was a 35-horsepower 8-cylinder JAP, not a Green. JAPs were made by John A. Prestwich, a former partner of Roe's, and as with all of Roe's planes, the engine was placed in front. I liked them this way since there was always the danger of the weight of the engine falling onto the pilot in a crash when placed behind him. I'd already seen how Roe's life had been saved by having the engine in front. The new triplane, unlike the others, did not have the three-plane lifting tail but a single float tail which made quite a difference to the feel of the triplane, as I soon discovered. It was like balancing on a knife edge.

Early one evening I wheeled the machine from the hangar for my first flight on it and was soon in the air. Some holes had been drilled around the cylinders to get a freer exhaust and at the bottom of the firing stroke flames spat out of them as well as from the exhaust ports. Oil came out of the crankcase and I got smothered. Worst of all was the position of the carburettor placed in the middle of the cylinders where the flames played around them. I was just about getting used to all this and to the new sensitivity of the plane when I realised I was at the end of the flying field with the railway not far ahead. Since I had gone right out of the flying area, I had to either land anywhere or try that more advanced evolution, a turn. Rather than damage the machine by landing on rough ground, I decided to risk a turn. Somewhat in fear and trepidation, I managed to bring the machine slowly round and sighed with relief when I was again facing the flying field. As I drew closer to the sheds, I could see everyone was out watching me. No doubt word had got around that I was flying out of Brooklands with the triplane and there they were, all waiting for something to happen. And something did. I came in to land. Easier said than done, I pushed the control column forward, too far. The machine dived and the ground rushed up to me at an alarming speed, so I pulled the column back and grabbed the throttle to open up as much as possible and shot up in a sickening fashion. Just in time to prevent a stall, I got the nose down again. Overdid things and again the machine shot up and nearly stalled. This seesaw business really worried me. I tried desperately to achieve a happy medium, believing it impossible, but eventually the machine began to settle. By now, however I was over the other end of the track heading towards Byfleet, so I had to make another turn. This time I made a very wide one. All went well, and I discovered it was now an easy matter to land, having acquired a forced degree of skill and confidence. I put the machine neatly in front of the sheds.

I eased myself out of the pilot's seat and walked towards our hangar feeling rather puzzled. There was not a soul in sight. The whole of Brooklands it seemed, had been watching me and I couldn't understand why the place had become suddenly deserted. Whatever had happened? Then Platt emerged, 'You should have heard A.V. carrying on.' But where was everyone? They thought I was a goner. My dives were just above their heads, I had scraped over the sheds without an inch to spare. Everyone had bolted into them and judging by the descriptions of my flight, I had been near death many times. *The Aero* reported it, 'The thrills of the evening were provided by Mr Pixton'. And they went on to explain, 'Mr Pixton had made a few straight flights before Mr Roe's visit to America, but this was his first attempt since then.' I had really upset Roe, but the storm soon blew over, and shortly after this irregular flight he recognised me as a fully-fledged pilot, apparently pleased with my work, and agreed that I should receive payment of £2 a week.

From thenceforth the money problems I'd endured working as a mechanic without pay were relieved a great deal and I was allowed to fly freely whenever I wished. I took the triplane out as often as I could to gain as much flying experience as possible. I got on famously but far-exceeded the £30 deposit I'd paid H.V. Roe for damage I did to the machine as I kept on crashing into the sewage farm inside the motor track on nearly every other flight. I would take off well enough, sail in the air with ease, start to turn and because of increased drag, the engine overheated and petered out over the sewage farm. The machine would sink gently and gradually under me as I fought vainly to hold it up, and we then had the distasteful job of squelching in the sewage beds to get it home for repairs. My unhappy landings were observed and reported. 'Mr Pixton had bad luck on the Avroplane, landing inside the sewage farm owing to engine trouble.' Others landed in the beds as well, but I did it so frequently that it became the thing to watch me. Whenever I wheeled the plane out for a flight, word (bush telegraph) went around, '*The Tripe Hound's out.*' All stopped to see the entertainment and, sure enough, I was back in the sewage beds and failed to get much sympathy. They would say with relish, '*Picky in the sewage again.*' Possibly because of my frequent encounters with the sewage beds, I gained something of a reputation for daring and earned the title of 'Sewage Farm King'.

On one occasion, when I'd successfully avoided the sewage beds, I was coming in from Byfleet to make a normal landing near the sheds but suddenly noticed a bump in the ground and elevated to clear it and was unable to stop before reaching the beds. I ran straight into them and stood the machine on its nose. A.V. really did love his triplanes and just hated to see me crash, but he soon got used to the crashes. I told him the trouble was the JAP runs for no more than ten minutes or so, then I'm down. I never sustained an injury, thanks to the low flying speed and the engine in front. Besides, the sewage farm was the softest part of Brooklands and further reports, often humorous, were appearing in the flight magazines. 'Mr Pixton, on the *Avroplane*, owing to the engine failing, once more found the attraction of the sewage farm too much for him, but he landed lightly and escaped with only a broken skid'. I'm sure the writer had his tongue in his cheek with another report implying that I couldn't stay in the air long enough to think of duration flying:

> 'When Mr Cody flying his Cody biplane on Laffan's Plain recently captured the all British record for distance and duration, 94 miles in 2 hours 24 minutes, the British airmen at Brooklands were determined to go one better. There were several aspirants to the honour, Mr Sopwith on the Howard Wright biplane and Mr Pixton on the *Avroplane*, being warm favourites. The odds however were slightly against Mr Pixton as the magnetism of the sewage farm had to be discounted.'

My flying was apparently being closely watched by a *Flight* representative who became very concerned for me:

> 'Mr Pixton is a very daring and pretty flyer, but the sudden movements he makes must put a severe strain on the bodywork. In particular one dive and sudden righting appeared to actually bend the body, and it speaks well for the work Mr Roe has put into his fuselage, that nothing serious happened. We do not wish to appear pessimistic, but if Mr Pixton continues his progress on his present lines, we doubt whether he will get through life unmaimed, and the science may lose the fine work which so promising an airman can put in.'

Flying produced a false sense of security. I never felt any sense of danger, and when I crashed it seemed as though I were stationary and the ground at fault, coming up to meet me. Even so, each time a crash was imminent I wondered, 'What will this one feel like?' We flew by instinct. We had to. Each of us was learning about the element we flew in, we learnt the feel of air vibrations, the breezes, gusts, air pockets and the seasons. Right-hand turns were dreaded by many as it was thought they were more difficult than left-hand ones, but this was merely hearsay and disproved technically. We did no flying in bad weather or in high winds, our machines could not stand up to anything but calm conditions. Roe had a little flag over the shed and by its fluttering we could see the strength of wind and decide whether or not it was suitable to take out a machine. At all times one had to be smart to prevent a stall. As I watched others crash and flew more, I learnt never to put undue strain on my machine, although I'd already been accused of doing so, for stalling was the easiest thing to do. Crashes were not usually serious as machines were light and had large surfaces and small engines. A stall meant that we dropped from the sky with a fluttering movement, like a leaf falling from a tree, and at the most suffered nothing more than a shaking up, a few bruises, seldom broken bones, but errors of judgement, however, accounted for crashes galore.

In early November I applied for my Flying Licence and felt confident enough to pass. Just before my tests I was up about 200ft practising figures of eight when the engine overheated again, there was the usual cut out and the usual sequel, the sewage farm... nose first! Oxley and the others who'd been running from the sheds to my aid, much to my astonishment turned back, reappearing with fire extinguishers. I'd not realised flames were behind me where the petrol tank was situated. There was a little hole in the top tank to let the air in, and when the plane was tilted up like this, the petrol overflowed and had caught fire, but I was able to snuff it out at source by putting my cap on it before the flames got a hold, so saving the boys a job. When they reached me there was nothing to do. So much for my tests to get a licence. The Aero Club tests were not difficult, but it was essential to make flights of about ten minutes duration and this engine of ours would not function long enough. The official who was due to watch me had to be put off till the next time, while A.V. was extremely sorry I'd been let down by the engine.

From the beginning the Avro planes suffered seriously from lack of power, otherwise they would have been very successful. A.V. had never done any long flights simply because of this. His triplanes, none of which were left apart from the current one we were using and the Harvard one I'd built up, had been fitted with 35-horsepower Greens, an excellent engine but with just not enough power, and now this JAP lacked the essentials for long distance flights. By good fortune Roe had passed his own tests with a Green. Sopwith was more successful than me, we were often in the air together, both eager to obtain our licences, but he got his on the Howard Wright biplane powered by a 60-horsepower engine around the time I had crashed and caught fire. His licence was No 31 but I had to wait hoping a promised engine would arrive soon to allow me to get my licence.

France had already issued more licences than us. They were issuing them in 1909, whereas we had only started this year 1910. France, in taking the lead, had firmly stamped her mark on aviation and several French words were with us to stay, *aileron, longeron, nacelle, fuselage, hangar* were some, *vol-plane* and *vol-pique* describing degrees of dives, *remous* the so called 'hole in the air', but Gilmour brought a new word into Brooklands, a word derived from London's new motor vehicle, the taxi-meter-motor cabriolets. Instead of *rolling out* a machine, we now *taxied* along the ground. Going were the expressions *flying machine, intrepid birdmen, aerial flight, chauffeurs* and *drivers* of aeroplanes, propellers were still being called *wind sticks* and *screws* by a few, the control column more than often was *stick* and to receive an air licence was *to get one's ticket*, but many still preferred the French expression, *getting one's brevet*.

There were no professionally run schools yet established in Britain, and those existing were all small and often one-plane schools. Unfortunately, aviation was a highly successful medium for a new type of confidence trickster. Not only were there the individuals who fancied themselves as designers, obtaining financial backing and producing something that would never fly or failing to produce a machine at all, but there were those who set up schools and advertised instruction for £100. Some of these self-styled schools were downright fraudulent but people anxious to fly would answer their ads. They would pay their £100 and be put in a machine incapable of flying for more than a few minutes, or even in machines not capable of flying at all. Some would never have left the ground in the hands of the best pilots in the world and, discouraged, the pupil would leave forsaking his £100. A few cases were brought to light where people, advertising themselves as a school, had no aeroplane. With pupils enrolling at £100 apiece, a machine was eventually bought and after several weeks' delay, the school called the pupils who were taught by inexperienced instructors on the very machine bought by their fees.

We started our one-plane flying school with Oxley as our first pupil. Roe accepted me as an accomplished pilot and was leaving all the flying side of the business to me. He was a brilliant designer, but not very good as a pilot, and was spending more of his time thinking up new ideas. Roe had given me no instruction. I'd taken up the old triplane on my own accord against his wishes on the first day I sat at the controls, so there'd not been much left that he could tell me. Therefore I was very much self-taught, and since I knew how to fly I was in a position to instruct even though I did not have an air licence.

We were attracting a number of fine men to the Avro School, men like Bell, Noel, Parke and Jenkins. Oxley, whom no one could mistake being anything but Yorkshire, had quickly become loved by the flying people. With his hearty laughter and fun-loving nature he proved quite a diversion, but he'd come to Brooklands with the idea that flying was easy, and each time he saw a crash he'd shake his head saying, '*Ee, they've no idea... Ee, what a to-do.*' I don't know how many times he must have said that when he saw me crash, but he changed his tune when he started flying as he found that he had no idea too, but he was made of the right stuff and had the makings of a fine pilot.

Instruction was verbal. As the School Instructor, I started pupils off on their first lessons telling them that they should only roll the machine along the ground at first, until they got the feel of the controls and hoped they would not take off as I'd done when Roe had trusted me with a machine. I would then take them up on a few flights so that they became accustomed to the sensations of flying. The controls and seating were so placed that the pupil could not take over. Whoever sat in the pilot's seat was the one in charge. Pupils, therefore, were more or less teaching themselves. Once they'd got the feel of the plane, I let them do solo flights at about 20ft, just straight hops, and when they had mastered the straights, I put them on left- and right-hand turns and figures of eight in readiness for their Aero Club tests. But we had our problems. One required infinite patience to learn to fly at the Avro School. Lessons had to be short because we still had trouble with the engine overheating and crashes were frequent, broken wings, broken chassis, broken wheels, but possibly because of much practice, we in the Avro shed had a reputation for quick repairs. A machine could be damaged one day and flying the next, but the Avro triplane was not the best thing in the world to learn on, it was difficult to handle, but safe. C.G. Grey believed it was possible to build a safe aeroplane, and had referred to our triplane during a discussion on the safety of aeroplanes as being a notable example. He reported, 'The Roe triplane has had more knocking about at the hands of pupils than any other machine, has been disintegrated for the *nth* time by Mr Pixton this afternoon, and yet nobody had ever been hurt on it.' Our greatest problem was that we were getting far too many pupils for a school with only one machine, and the weather was often far from suitable for flying being winter, which didn't help matters.

Bristol, the British and Colonial Aeroplane Company, now with two sheds at Brooklands, had started a flying school under the charge of Low. They also had one at Larkhill, Salisbury Plain, having leased a large area of land from the War Office. As the company's intention was large-scale production, they worked on the principle that they would not sell planes unless there were pilots to fly them, and went all out to build up reputable flying schools. Another one-plane school like ours was the Blondeau Hewlett School. Blondeau had lived in England for many years and learned to fly at the Farman School in France, then he settled at Brooklands where he formed a partnership with Mrs Hewlett, who was the senior partner, and between them they ran the school with one Farman. Mrs Hewlett, daughter of a vicar and known in motoring circles, was married to the well-known author Maurice Hewlett. She attracted attention at Brooklands, not only by being the only woman amongst a crowd of men, but by usually wearing clogs, and very wise too for it could be extremely muddy after a downpour. She did not know how to fly yet, but was learning and hoped to get through her tests soon.

There was another Frenchman in the hangar next to us, Maurice Ducrocq, also with a Farman. He'd been taught at the Blondeau Hewlett School, and had received licence No 23. Ducrocq lived nearby, lodging with a painter and his wife. Gordon England had also formed a one-plane school, the Hanriot School, when he discovered a machine, a French made Hanriot, had been left in a hangar and was for sale. He bought it and called it *Henrietta*, and with it so started his flying school. We therefore had at Brooklands the Avro School, the Bristol School, the Blondeau Hewlett School and the Hanriot School, all in their infancy and friends teaching friends. It was only at the end of 1910 that serious flying schools started up, with Bristol's becoming the most thriving school of them all.

As 1910 came to a close, serious flying was also getting underway. Great sums of money given as prizes by the *Daily Mail* had encouraged the first spectacular flights in Britain, but a private man had made a wonderful offer. It was Baron de Forest, a British subject resident in England since the beginning of the century, owner of large estates abroad and a Member of Parliament who, unlike many of his fellow members, showed a great interest in aviation. Earlier in the year he'd offered £4,000 to the person to fly the longest non-stop flight from England to the continent, a contest for British competitors only. Over forty entries were received by the Aero Club, including entries from Roe and myself, from Ogilvie, Sopwith, Moore-Brabazon, Grahame-White, Cody, Gilmour, Grace, Loraine, Macfie, McClean, Barber, Neale, Barnes, Morison, Fisher and many more. Although the Baron de Forest competition was made known earlier in the year, no one attempted a flight until December when the number of active competitors had shrunk considerably. Reluctantly we were forced to withdraw as our engine had not been replaced, the old one would certainly let us down before we'd gone very far.

Of the active competitors there were Grahame-White, Loraine and Greswell who went to Swingate Downs near Dover waiting for favourable weather. McClean, Sopwith and Grace were waiting at Eastchurch, Cody was hoping to start from Farnborough, and Ogilvie hoping to leave from Camber Sands. Favourite to win was Grahame-White. The days went by and the competitors weren't doing very much in the way of flying. Grahame-White met up with several difficulties. He had a new Bristol biplane for the flight, but during one of several test runs he crashed when a gust caught him which resulted in an injury to his leg, cuts to the face, and his passing out from loss of blood which put paid to further flights and hopes of winning.

Bad weather tore down Loraine's and Greswell's sheds and badly damaged their machines but time was running out. Cody hadn't started, Ogilvie attempted a flight but decided that the wind was unfavourable, in fact the December weather was against all the flyers. The first person to do anything worth noting in the contest was the very popular flyer Cecil Grace, who set off from Dover, crossed the Channel and landed at Les Baraques near Calais, but as he felt that this distance would be beaten by another competitor he decided to return to have another try. He was never seen again. For days we hoped he'd been picked up by a boat, but no. His goggles, identified by a friend, were washed ashore on the Belgium coast, his body never found, he was assumed dead. Cecil Grace was the second British pilot to die.

In the meantime, Tommy Sopwith slipped off early one morning, crossed the Channel safely, battled through fog and landed after flying for three and a half hours not knowing where he was. He was, however in Belgium at Thirimont where two witnesses verified his exact point of descent. He'd flown a glorious 169 miles, and as the Baron de Forest competition had no further starters, Sopwith had won £4,000 prize money. *Flight* was just as pleased as we were:

> 'The achievement of our man Sopwith is the sole topic of conversation in the Blue Bird. We thought he would get across the Channel, but to fly 169 miles in the first cross-country flight he had made surprised our wildest expectations.'

Another incentive to the British airman came from the Michelin Tyre Company who offered £500 for the following five years for the longest flight of the year. With it went the Michelin Cup. Although Cody made no mark in the Baron de Forest contest, he'd been setting up duration records competing for the cup with Sopwith as a strong contestant. During November he had flown 94 miles at Laffan's Plain, Sopwith 107 miles at Brooklands, but both improved their records in December with Cody flying 114 miles non-stop flying and Sopwith 150 miles. Ogilvie joined in and flew 142 miles at Camber, but Cody finally took the cup after flying round and round Laffan's Plain for over four and a half hours on 31 December. He came down joyful but cold, having set a new British record of 185 miles non-stop flying, the longest flight ever made in Britain – a great achievement, and a greater distance than Tommy had flown in the Baron de Forest contest. The Michelin Cup, a trophy of a winged horse rearing, and a man seated on its back holding the reins in one hand, was presented to Cody. On it was inscribed *'The man, by the aid of his aeroplane, overcomes the attraction of the earth'*.

At last Roe secured another engine, a 35-horsepower, water-cooled Green, which was fitted into the triplane and I applied for my air tests again. Watched by Mr Perrin, I passed easily gaining air certificate No 50 which was granted under these rules:

'Three separate flights must be made, each of three miles round a circular course without coming to the ground. These flights need not necessarily be made on the same day. On the completion of each flight, the engine must be stopped in the air, and a landing effected within 150yds of a given spot previously designated by the candidate to the Official Observers.'

Oscar Morison, James Valentine, Astley, Robert Macfie, Victor Sassoon and I all passed at Brooklands on practically the same day. Now we were members of the Aero Club we received our copies of *Flight* free of charge. A.V. and I were the only two in Britain to secure licences on a triplane, and though records show I got mine on the 24 January 1911. The date on my licence is 31 December 1910, the last day of the year.

The year 1910 was when flying at last got underway in Britain and on the personal side I'd not done badly since June 1910. I'd been connected with flying only six months and had attended an aviation meeting, been to America, came to know most of the pilots in Britain, got my Flying Licence and was a flying instructor. What would 1911 bring? Although Cody had put up a British duration flight of over four and half hours and Sopwith had flown a three and half hour non-stop flight to the continent, exceptional flights, the public were of mixed feelings where pilots were concerned, and the following witty piece in the *London Opinion* sums up 1910 as perhaps nothing else could in so few words.

In the air 1 minute.	Another foolish Aviator!
In the air 3 minutes.	Hasn't he killed himself yet?
In the air 5 minutes.	All the fools ain't dead yet!
In the air 30 minutes.	Mr Ayrider, the well-known Aviator!
In the air 1 hour.	Our distinguished fellow-countryman!
In the air 1¼ hours.	The Wizard in the Air!
In the air 1½ hours.	The Legion of Honour could have been bestowed on no worthier man!

CHAPTER FIVE

With Roe Flying the Avro Biplane

A.V. Roe and I had become very friendly. He invited me to his home many times and on one such occasion at the beginning of 1911 revealed that he had second thoughts about triplanes. He'd given up the idea of the triplane being the best form of aeroplane and showed me his drawings of a new biplane he wanted me to test. Roe had won several prizes as a racing cyclist and had gained his engineering knowledge with a railway company and while at sea as a marine engineer. He'd been thinking a lot about flight when he was about 25-years-old, just before the Wright Brothers flew. In fact, he'd cycled from England to France where he met Wilbur and was very open about his ideas of flight. He also wrote to *The Times* but they weren't particularly sympathetic to his ideas. He had built a pusher biplane with the propeller behind its wings before his triplanes, but was not taken seriously until after Blériot's flight across the Channel in 1909. It was not easy in those earlier days.

During January 1911, a young man of nineteen rode up to Brooklands on his motorcycle, his first time at the ground, his name... John Alcock. He had an engine for Avro from the Empress Motor Company. We had been expecting it. Young Alcock was very much taken with Brooklands, so impressed was he that he wondered whether there was a chance of getting work here and I introduced him to Ducrocq next to us. He succeeded in persuading Ducrocq to take him on and was overjoyed at being accepted. I tried out the Empress engine he'd brought but it did not prove much good in the triplane, not behaving well at all, but while I was tinkering with it, Macfie wandered over. He was interested and examined it, tried it on his machine and liked it.

Meanwhile the Government had not been over-enthusiastic about the development of aviation, and while French and German Governments were financing and encouraging flight, private fortunes had been spent in England. Captain Dickson, who'd been involved in the first midair collision, had done much in trying to convince the Government on the future of the aeroplane for the Services, stressing its advantages for air reconnaissance. His observations and ideas were brushed aside with a short-sighted comment, *'Those playthings will be of no use in War'*. The Government was playing safe and a cartoon appeared in a newspaper of an aeroplane flying into a brick wall, an attempt to illustrate the Government's attitude towards aviation.

At last by February 1911, the Government was sufficiently impressed by the potential of the aeroplane for military purposes that it issued particulars of the formation of an Air Battalion. Naval and military men were encouraged to learn to fly privately and be reimbursed on gaining a pilot's licence. They could then transfer to the Air Battalion on application. It was not considered necessary to lay down definite rules, but the following qualifications would be taken into consideration:

Special recommendation by commanding officer, possession of Aviator's Certificate, previous experience of medical fitness for air work, previous experience of Aeronautics, rank not above captain, good eyesight, good map reader, good field sketcher, unmarried, not less than two years' service, under thirty years of age, good sailor, knowledge of foreign languages, taste for mechanics, light weight, under 11 stones 7 pounds.

The formation of the Air Battalion was a great step in British aviation. But France had 344 trained pilots at the end of 1910 while Britain had fifty and Germany about the same, with Germany specialising in the study and production of airships. Bristol was quick to realise that the company could expand its flying schools by taking in servicemen and offered tuition at special reduced rates of £75. It was in an excellent position at Salisbury Plain since the Army had several camps in the area, therefore men wishing to fly did not have to seek out a flying school miles away from their bases. Bristol advertised extensively. 'The British & Colonial Aeroplane Company Ltd. Flying Schools at Salisbury Plain and Brooklands. Special Terms to Navy and Army.'

A new flying ground had recently been open to experimenters at Hendon on the spot where Paulhan had taken off at the start of his famous London to Manchester flight. It was being run by the London Aerodrome Company and its representatives emphasised that the ground would be for regular tenants only, that they would not be promoting it into anything like Brooklands where crowds came to watch flying. This was not to be. One of the first tenants was Blériot's manager, Monsieur Chereau, who'd been the London manager of Blériot's motor lamp business and who had taken charge of things this side of the Channel during the famous Channel crossing. Now he was in charge of the Blériot Aerodrome Company flying school and aeroplane works at

Hendon and with him was Monsieur Prier, the school's flying instructor.

Then Grahame-White took over Hendon, he set up a new company with Blériot and Sir Hiram Maxim and had his own ideas about Hendon's future. He intended to run several flying schools, erect more hangars and arrange flying meetings. The ground, consisting of 270 acres, was bigger than Brooklands and he announced that Hendon could be open to the public by Easter this year. He opened his own school and J.V. Martin, the Harvard University's representative who'd invited Grahame-White and Roe to America, was in England again and learnt to fly there in next-to-no time, gaining the British Flying Licence, No 55.

Around this time, we were delighted to hear that Tommy Sopwith had received an invitation from King George V to fly at Windsor, but when he left Brooklands to keep the appointment it was foggy and showed no signs of improving. We learnt that he'd had a bit of engine trouble on the way, which he put right, and luckily as he neared Windsor the fog lifted. He alighted on the Royal golf links and the King greeted him, then before returning to Brooklands, Sopwith took tea with the Royal Family.

Seldom did we see a freak machine now. Up to this time there had been quite a spate of weird aeroplanes which were rapidly dying out, their inventors vanishing from the scene. Designers generally stuck to accepted principles, producing the pusher biplane or tractor monoplane but Passat, a Frenchman living in England, closely followed the flight of the bird and had conjured up the *Passat Ornithopter* which he brought to Brooklands. He swore that he'd been able to raise it off the ground, but I never saw it move an inch upwards and by the nature of its design I was certain it would be impossible. It was the strangest machine I'd ever seen, looking like a small airship with flapping wings. Roe came out to have a look at it. Passat was one of the many to whom I'd written when I was trying to find someone in the flying world to take me on, his negative reply was to my advantage. Most people had long realised the flapping-wings principle was a thing of the past and could never be the answer to flight, but occasionally there were those with strong commitments based on observations of birds in flight saying they had studied birds and insects, with some still believing that man has all the power in his limbs necessary for flight.

Another weird machine attempting flights at Brooklands was the *Rhomboidal*, a diamond-shaped freak with a 30-horsepower Humber engine, whose whole structure was flexible instead of being rigid, its main spars composed of stranded steel wire. Work came to a stop, as it always did when something unusual was going on, as it made one of its notorious efforts to fly. On this particular day we watched it travel 50yds or so along the ground at a little over walking pace, as it flapped its wires trying desperately to rise, shuddering and shaking in the process, and after a while it quietly sat down and crumbled. That was the last thing it did, but its inventor was in no way discouraged by its impractical construction, convinced he would get it flying yet.

There was a handsome triplane, its owners were associated with the Molesworth Engineering Dictionary, which had been at the ground for a considerable time. It was called *Britannia*. The wings were of equal length the lowest one nearest to the nacelle, the other two staggered towards the tail which gave it the appearance of steps. Because of this we nicknamed it the *Flying Staircase*. It had been reported as being, 'several times on the point of lifting'. I didn't see it fly and, in common with many, I don't think it could, and it was eventually taken away from Brooklands. There was another freak machine I saw, but not at Brooklands, designed by a man called Walter Windham, a machine that looked like a telegraph pole and when he posed on it to have his photograph taken it collapsed.

Hammond, who'd experimented with helicopters, turned to designing a triplane on which he'd been working for some time in the shed constructed from tea chests. The machine was not large, but because of its homemade appearance everybody took a deep interest in it, especially me when I was asked to fly it. Unlike Roe's triplane, the *Hammond* was more a copy of the pusher Box Kite with three pairs of wings. All activities came to a standstill when it was brought out. The metal work was a mass of holes, laboriously drilled out by hand to lighten the machine, and as I climbed in, a cheer went up. Some wits lay flat on the ground to see whether they could perceive a gap between the wheels and earth. Of course, no one expected it to go far except perhaps Hammond himself. As I opened the throttle, it lumbered forward and I felt the front wheels just lift clear, that was all. Its little V-engine gave no power, but this effort alone proved too great a shock for it backfired and caught fire. I performed some smart work around the carburettor, again with my cap which was proving most useful for small fires, but the damage was of such an extent that it put an end to the triplane's career.

One of the main difficulties constructors faced was to make their aeroplane stable in flight. I was speaking to a man who had seen an aeroplane for the first time and he thought it was flapping its wings. He was not to know that he'd only seen a Box Kite rolling from side to side due to its instability.

About this time Roe received a letter from a man in Winchester, he wanted a pilot to test the *Hawkin* machine fitted with an automatic device for lateral stability. A.V. asked if I would try it out for him. I agreed and on arriving at Winchester I was met by the inventor, Mr Hawkins, the aeroplane was in a shed about 3 miles away. We drove to the shed. The ailerons on each wing were connected by a bar on a pivot placed above the pilot's head and worked like a seesaw. Hawkins explained the idea that when a gust hits one wing it would lift up the aileron on that side, and on doing so the bar will push the other aileron down, thereby keeping the machine on even keel – simple enough in theory. Roe's machines had ailerons which we pulled up and down with the aid of large levers at our side, but I had my doubts about this device and thought I should have some control over it. We were out early next morning, and I started up the little Alvaston engine, taxied to the far end of the field, opened out and rose. Then the fun commenced. Instead of keeping the machine on an even keel, as the designer had hoped, the ailerons banged up and down at great speed, rocking the machine so violently that my teeth rattled. Needless to say, I closed the throttle as soon as I could get my shattered wits together and got down more by good fortune than judgement. When eventually I pulled myself together, I bade Mr Hawkins a cordial farewell and made rapid tracks for the railway station and the comparative peace of Brooklands.

One day during March having successfully flown the triplane in a high wind and landed, I decided to go up again and was turning the machine quickly around to start, when the body broke off behind me, but we soon fixed that so it was ready to fly again in next to no time. I was thankful it hadn't happened while I was actually flying.

During the same month the new biplane about which Roe had spoken arrived from Manchester and we soon had it assembled and ready for tests. It had a triangular fuselage, two large landing skids, weight 500lbs, and painted in large letters under the wings was AVRO. It looked good and was a tandem plane, the pilot sat behind the passenger. Roe favoured the propeller in front and had pioneered the tractor-type of aeroplane in Britain with his triplanes. Most other machines were either tractor monoplanes or pusher biplanes of the box kite style. This was Britain's first successful tractor biplane and I was the first to fly it... I got in it, it rose and handled beautifully. I was delighted and felt that this was going to be A.V.'s first real success. Still only powered with a 35-horsepower Green water-cooled engine, it averaged 45 miles per hour, and as I landed at high speed A.V. came towards me to hear the results of the tests. It was an outstanding machine, a pleasure to fly, stable, very easy to control and a vast improvement on the triplane. 'In my opinion, it must be the best all-British machine in existence.' He looked very pleased, and said he would enter me for the forthcoming Brooklands to Brighton Race.

A.V. took my word regarding the plane's performance and did not even want to try it for himself. In fact, he'd not flown for many months now and had perhaps decided not to fly again. A report on the appearance of the Roe biplane at Brooklands ran:

'Pixton took it out for the first time in a regular gale, and flew the length of the ground in most exciting style, the machine showing excellent stability. So strong was the wind that, when he came down at the paddock end of the ground, Pixton could not get the machine round because his rudder was always blown down-wind.'

Very quickly the little machine made a name for itself, being referred to as *'That wonderful little Roe biplane'*. A.V. paid me the compliment of entrusting his wife with me for a flight, and as I became so fond of the biplane and was constantly flying it, it was being called, *'The Pixie Plane'*. Another report appeared, 'The success of the new biplane has been one of the most striking features of the Brooklands flying ground for some time.' Roe's prices for his planes were, and always had been, very fair and the company now advertised the new biplane:

'A.V. Roe & Co. Britain's most successful aeroplane is unquestionably the Avro Biplane. £600. Everything halved but efficiently. Small upkeep. Small size. Small hp. Its speed, small size and smart appearance being specially noticed by spectators.'

The Brooklands racing season started at Easter and would continue to October. This meant flying meetings would be held once or twice a month. There was record attendance on race days now, the crowds more than doubled by the added attraction of the flying, but we still had people standing around on non-race days since Brooklands was considered to be one of the most interesting places to visit in Britain and had become a regular spot for motorists and their friends at weekends. Messrs Keith & Prowse, the well-known box office proprietors, were booking passenger flights by arrangement with the pilots and had erected a little office at Brooklands, near the Blue Bird Café, equipped with a telephone so that their branches had a direct line of communication with us. At a minute's notice they would know whether flights could be arranged at a given time. They were our sole agents and did a fine job as middlemen, having cars to meet their clients to take them to the shed and pilot of their choice. Men were not our only eager passengers. We took up many a daring lady and never once did I have a frightened customer. It became quite the thing to have photographs taken beside an aeroplane, especially with the pilot, and autographs were in great demand. Bell, who began his flying at our school, once played the fool with a passenger by flying a few feet above the ground and around the sheds, scaring the senses out of the poor fellow which, I thought, was rather an unkind thing to do, but this type of incident was fortunately very rare. On the whole the public thoroughly enjoyed themselves at Brooklands, feeling they were in touch with the modern world.

The Royal Automobile Club was now offering prizes for competitive flying. No flights under fifteen minutes counted, but three prizes were to be given for the longest flights, £30, £15 and £5 and an aggregate competition was arranged for the pilots who put in most flying during the season with awards of £150, £100 and £50, and I thought I had a good chance of getting some of that money. Brooklands was becoming sophisticated. Our names were posted up on large boards which fitted into a frame so that the spectators could follow the flying whenever someone left the ground. George Reynolds became the very well known figure on top of a hangar as he followed the proceedings and made the announcements from a chair in the capacity of timekeeper. Ebblewhite was another timekeeper, but I didn't get to know him as well as I did Reynolds who used to fly with me for pleasure.

I was in the air as much as possible and soon found I was winning all the first prizes for being in the air the longest. I welcomed the money but, above all, I loved flying. I confess, too, that I enjoyed the publicity which followed, accounts of my flying at this time were plentiful and ran like this:

> 'The flying at Brooklands was limited to a couple of lengthy flights by Mr Pixton on the Avro biplane... Mr Pixton was the only one to venture aloft... Pixton had the field to himself... Pixton was first out... Mr Pixton's was the star performance... he thoroughly deserves the £30 he's won... Mr Pixton, as usual with his disregard for the elements, gave a clever display... In spite of a stiff wind, Pixton was determined to try a show with the Avro biplane... Mr Pixton was the man so far as flying was concerned, the breeze freshened, he still went on, some of his turns were watched with no little anxiety... Pixton's flying was admirable in every way... It was blowing the best part of a gale all afternoon but Pixton made a magnificent flight of 55 minutes... The aeroplane pitched about like a small boat on a choppy sea, but Pixton seemed to rather enjoy it....'

Then the Green Engine Company started using our name for advertising, the following advertisement appearing in the May issue of *Aero*:

> 'Green's All British Aero Engines. Brookland's Easter Meeting 1911. Pixton flying the Roe Biplane fitted with 30/35 hp Green Engine flew for 87 minutes 37 seconds in a very strong wind. It was the best flight of the day, and won the Brooklands Endurance Competition, creating at the same time a record for 30/35 hp engines.'

After the race days we returned to the normal routine of flying, maintenance and experimenting. Since aviators and motor drivers didn't mix on the whole, the clubhouse in the paddock where the cars assembled was used only by the motorists, while we collected together at the Blue Bird Café. In the evenings many of us frequented the King's Head in Byfleet.

Although I crashed the triplane I never crashed the biplane, nor sank into the sewage farm with it, but it did not escape damage in the hands of other people. Conway Jenkins, ex-motor salesman, had just secured his licence at the Avro School and he took the biplane out with his first paying passenger just a few days before I was to compete in the Brooklands to Brighton Race. Still rather inexperienced but full of confidence, he took off with his passenger climbing much too steeply, rose to about 100ft, tried to turn and stalled. The machine came fluttering back to earth and finished up on its nose. I was shocked to see it come to grief so soon. Nothing happened to Jenkins, he escaped with hardly a mark, but the passenger's face had been forced into the radiator grill in front of him, and a journalist with a warped sense of humour wrote afterwards that the passenger had alighted '*suffering from filleted face*'. The unfortunate man was not called upon to pay for his flight, but alas for A.V's plane, it had been damaged so seriously that it had to be completely rebuilt. Luckily, the Brooklands to Brighton event was postponed because of unfavourable winds and we were able to have the biplane ready for the race. Then, unbelievingly, I watched as another pupil put it into the River Wey, but it was none the worse for this adventure. From then on I made sure that no one came anywhere near the biplane until the race was over, the pupils had to be content with the use of the triplane only.

Eventually, the Brooklands to Brighton Race was held on Saturday, 6 May 1911 after three postponements. It aroused immense interest since it was the first point-to-point cross-country race to be held in Britain, total distance about 40 miles. Headlines appeared, '*Great Air Race, Brooklands to Brighton.*' A few days before the start, Gordon England approached me. His father owned a large country house at Haywards Heath, called *Oakwood*. I'd never been there, but I believe it was a beautiful place, and the family were well liked and respected. They were arranging a flying display for the residents and asked me would I fly for them on my return trip; he and Morison would be flying as well and his father would mark a field to land in with a large cross. I'd taught Gordon the rudiments of driving and he had only just gained his flying licence, No 68, at the Bristol School, his Hanriot Flying School having folded up. Geoff, his youngest brother, who was very keen to fly, was often at Brooklands.

The day of the race came, huge crowds assembled at Brooklands and thousands travelled to Brighton. Several pilots had put their names down for entry, including Greswell, Valentine and Morison, but it finally boiled down to four, Hamel, Snowden-Smith, Gilmour, and myself. All of us, except Hamel who flew in from Hendon, were Brooklands' men. As the first race in Britain it was a very important event in the history of aviation and I would have liked to have won it. The earlier flights had not been races, the London to Manchester, the Baron de Forest, the Michelin Cup, were all challenges. No one had lined up and left at a given signal, but unlike them the first prize was not in the region of thousands of pounds, nor even hundreds. It was £80, a small reward by comparison, but compared with my weekly wage of £2, it was quite a sum.

Mr Ebblewhite was ready in a new timekeeper's box at Brooklands but there was a last minute change. The race would be handicapped according to the type of machine, the first handicapping introduced into aviation in any part of the world. It was decided on the speeds of the machines, Gilmour would be off first, followed by Snowden-Smith four minutes later, myself next and last Hamel. Snowden-Smith's Farman was built by Mrs Hewlett and Blondeau, their first one, and they were hoping to extend and build more machines in the future. Hamel was expected to do well and my chances of winning on the popular biplane were considered good, but I lost the race before it had begun.

6 May 1911		
First Point to Point Race in Britain		
Pilot	Plane	Speed
GILMOUR	Bristol Box Kite	35 mph
SNOWDEN-SMITH	Farman Box Kite	40 mph
PIXTON	**Avro Biplane**	**45 mph**
HAMEL	Blériot monoplane	55 mph

I was putting in some time for the £500 Manville passenger-carrying prize, which I was hoping to win for the accumulative flying time on certain days between Easter and October, and this day was one. I was flying without a watch but an arrangement had been made to flag me down in time for the start of the race at 3pm. I was still flying when suddenly I realised the race had started. By the time I landed, topped up the petrol, checked the oil and water and got away, I was about fifteen minutes behind, but I was determined to have my trip to Brighton. I thought, though, that they might have waited for me, I was, after all, 25 per cent of the field. So, starting at a disadvantage, off I went. I met a little air turbulence as the land rose just south of Brooklands, but apart from that I was flying well and enjoying the flight over the countryside. I'd been flying for about forty minutes when I thought I should be near Brighton, but there was no sign of the town or indeed the coast, or any of the other competitors. With a sinking feeling I realised the petrol supplies were getting low, and that at any minute the engine was liable to cut out. I looked for a suitable stretch of ground, saw a green patch below me, and made a landing.

A crowd of spectators arrived on the scene from all directions, for the sight of a plane over the countryside, especially one coming down to land, was a great spectacle in 1911. 'Where am I?' It was Plumpton Race Course. I was about 7 miles from Brighton. Some wanted to jump in and show me the way, but I said I would be able to manage now. I'd failed to maintain an accurate course as the compass I had was a primitive thing and proved most unreliable. The vibration from the engine made the needle swing around in a useless fashion, but where could I get hold of some petrol? There was none at the racecourse and indeed I had great difficulty in finding any at all. Eventually I got some from Wells Garage, and as I felt hungry I went into a pub, the Sun Hotel, and had some tea there before restarting. By the time I'd finished and returned to the plane, the small crowd had swelled to handsome proportions. Probably no one in the crowd had ever seen an aeroplane and they were making the most of the occasion by writing their names all over it. One lady greatly embarrassed me by insisting on making me a presentation of a bunch of primroses.

I instructed half a dozen men in holding the aeroplane tail as I prepared to leave, explaining the importance of hanging on so that I could start the engine without the danger of making a premature lurch forward, as the machine would pull very strongly as soon as the engine got going, 'but don't let go until I give the signal by waving my hand,' I said. My helpers were eager but, concerned with the importance of their task they threw all their weight onto the tail at the first sound of the engine and all pulled downwards in one accord evidently thinking it would rise like a balloon. The rudder post, to which the tail skid was attached, couldn't stand up to this type of treatment and broke.

I was still determined to get to Brighton that day but as I could not fly without a rudder operating, I set to repairing the damage. Watched by a very keen audience, I took a stake from a nearby fence, shaped it with my penknife and tied it to the broken post with a bit of wire so that it acted as a splint.

With the crowd more excited than ever, I made a second attempt to leave taking greater care this time in impressing my amateur ground staff exactly what I wanted them to do. All went well and, with the bouquet of primroses tucked in my jacket, I waved to the cheering crowd from the safety of the air as the rudder functioned behind me as though nothing had happened.

The prom at Brighton had been packed with people, but there wasn't anyone there when I arrived at 6.20pm, over two hours after the winner. I passed the captive balloon marking the finishing line at the end of the Palace Pier, delighted to get there at all. I'd taken three hours. Hamel, the youngest man of the race, not long out of Cambridge University, had won on his Blériot. He'd covered the 40 miles in 57 minutes and took the £80 prize. The Lord Mayor had greeted him, and he had signed his autograph on all manner of bits of paper offered to him. Snowden-Smith was second on the Farman Box Kite having taken 1 hour 21 minutes to complete the journey, a considerably greater time than Hamel, but was disqualified as he missed a compulsory turning point, so Gilmour on the Bristol Box Kite officially took second place. Gilmour, however, waived his claim to the prize of £30 and took the third prize of £20 with £20 for being the first biplane to officially arrive in Brighton. But for my ridiculous late start, the Avro biplane might have equalled the Blériot. It was very regrettable that I'd been unable to give it a fair run.

I met up with the others, and we went to the Albion Hotel where we were the guests of Harry Preston, the owner and also an organiser of the race who gave us free accommodation for the next night or two. As well as being an ardent supporter of flying, Mr Preston had a passionate interest in boxing. Still with my bunch of primroses, I spent a most enjoyable evening in The Albion. Mr Preston laughed heartily at my adventures in Plumpton which seemed to tickle his particular sense of humour, with the rudder splint arousing a great deal of interest. The next day, newspapers and magazines told their stories.

> 'After three successive postponements, the great aeroplane race from Brooklands to Brighton passed in history yesterday, the first real picture of British aerial prowess... Not since the Manchester flight has an aviation event in this country stirred the public's imagination to the same extent as yesterday's contest. The spectacle of four young Englishmen making history in the first legitimate aerial Point-to-Point Race in history, brought huge crowds to Brooklands where the race was due to start at 3.00pm... There was the largest off-day crowd ever seen on the course to witness the start of this memorable race, which

can now safely be said to have been far more successful than the motor run from London to Brighton in 1896...'

And that was in the *Autocar*. The first two reports were in the *News of the World*, and there was much more. *The Aero* wrote of my late start:

'Pixton, the brilliant pilot of the little Roe Biplane, surpassed all expectations. Under the impression that the Brighton race started at 3.30pm, he went up not long before 3.00pm with a passenger, and flew for twenty-eight minutes in competition for the Manville passenger prize. When he came down he found that the others had started, so he simply filled up with petrol, oil and water, unloaded his passenger, and went off after the others.'

The *Sussex Daily News* told of the excitement at Plumpton:

'Plumpton racecourse enlarged its experience in racing matters on Saturday when it figured prominently in the aerial contest from Brooklands to Brighton. The excitement of many a spirited steeplechase over the course was hardly comparable with the enthusiasm which was aroused by Mr Pixton's brief visit. Farmhands laid aside their implements, the villagers left their gardening or other rustic pursuits, and all made a beeline for the race course to inspect in blank wonderment the mysterious craft which had swooped down upon them.'

It was Sunday, and I was still in Brighton. On Monday the prizes were to be presented by the Lady Mayoress of Brighton, but as I was receiving none I was free to keep my appointment at Haywards Heath earlier than I might have done otherwise, so I left my flying partners in their glory and set off. Haywards Heath was only about 12 miles away, so it didn't take me long to get there. From the air I saw Mr England's fine residence in a hollow surrounded by trees. Countless motor vehicles were dotted over the countryside, their owners looking skyward as I flew in, and the cross was there which Gordon had mentioned, easily distinguishable in a field, but I felt the field was a bit on the small side so came down in the adjoining one which had much barbed wire surrounding it. I assumed it must also belong to Mr England. Again the *Sussex Daily News* went to work:

'The district became quite excited... It was a lovely afternoon and hundreds of people visited Oakwood and awaited results. Hope was fast slipping away when the keen sighted caught a glimpse of an aeroplane coming from the direction of Brighton... the aviator was received with cheers and there was an immediate rush of people from all directions to welcome him and obtain a closer view of the machine, the first one to alight in mid-Sussex. The hedgerows were much guarded with barbed wire, and many people came to grief in the entanglements. There will be busy days for tailors this week. The visitor proved to be Mr Pixton, the hero of the previous day's flight over the district...'

The biplane collected more signatures alongside those scribbled by the people of Plumpton as Mr England greeted me, a charming man. He took me into the house and made me very welcome. Gordon and Morison had not arrived yet but he had received a message to expect them the next day. After having tea with the family, I gave a short flying display and invited a few people for a flight. They were delighted to have this opportunity to fly in an aeroplane. When Morison flew in with Gordon he landed about 2 miles away because he'd run out of petrol and there was an instant rush for motorcars and bicycles, but when they were seen flying towards us there was a rush back. It was a wonder no one was hurt. We gave a two-day demonstration and some passenger flights.

In between the flying, I spent some quiet moments with Geoff who took me around on his motorbike, a Bradbury's, as I sat snug in the sidecar with a rug wrapped around my legs, enjoying the country. When he was a little older Geoff intended joining his brother and making flying his career. I had kippers for breakfast during my stay. Apparently Gordon had told his father how I was fond of them and they'd especially gone out of their way to see that I got some. Then it was time to leave, but before I left I thought I'd better fit an entirely new rudder post on the biplane and made a really sound repair out of a broom handle. It was a good job done; it would probably last the rest of the machine's life.

The flying display had been a great success. The field I'd landed in on arriving, however, belonged to the nuns nearby and an amusing account of both my arrival and departure was written up.

'Messrs Pixton, Morison and England certainly roused Haywards Heath. Pixton literally outdid Coriolanus when, like an eagle in a dovecote, he fluttered the Sisters of the Holy Cross Home on his wonderful descent from Heaven amongst them. These ladies are a Church of England sisterhood, and one of them gave me a graphic account of the proceedings. It appears that the Mother Superior could not get her brood in to Vespers, as they were all fascinated by either Pixton or his machine. The sister's description of his departure was delightful. "They first of all twisted the propeller and wound the machine up, then they all had to hold it back, and when he put his hand up they let go, and the machine rose gracefully, and he flew off like a duck". I do not know whether the machine or Pixton was described by the last.'

But the sisters were not happy. The Mother Superior objected to the disturbances created by the flying and Mr England received a request from a lawyer acting on her behalf of a charge of £5 for the use of their field and cost of repairs to their hedges. The lawyer explained, 'I am informed that a large crowd assembled at Oakwood and that my client's hedges and fences were broken down, and that their private road was blocked up by motorcars and bicycles.' Mr England was also threatened with an injunction to stop him from organising further 'Aerial Exhibitions' if he were not able to give a definite assurance that there would be no future flying at Oakwood.

On leaving, Morison and Gordon were not so lucky as me. In rising from the hollow, Morison fouled the trees, sank beautifully into them, and had to be rescued by people who rushed off for ladders, but then he was always crashing. Not only did he land the aeroplane in England's trees, but he'd fallen in the sea at Folkestone, had been caught in electric wires on failing to notice them at Eastbourne, and had landed on the shingle at Brighton smashing his propeller.

The first major event to occur at Hendon was the Parliamentary Demonstration organised in mid-May by Grahame-White in conjunction with the Parliamentary Aerial Defence Committee headed by Arthur Lee and Arthur du Cros. There'd been scarcely any official air policy and the aim was to impress Government authorities of the role the aeroplane could play in aerial defence at the time of war. The Prime Minister, Mr Asquith, attended with most of his Cabinet and about 200 Members of Parliament and included Mr Balfour, (Leader of the Opposition), Mr McKenna (First Lord of the Admiralty), Viscount Haldane (Secretary of War), Colonel Seely (Under Secretary for War), Winston Churchill, Lloyd George, Lord Northcliffe and many other distinguished guests. I was one of the few invited to fly and A.V. came to watch. Cody, Loraine, Hamel, Drexel and Barber were others along with Grahame-White's men, Grahame-White himself taking a major part.

The demonstration opened with the inspection of machines, the dismantling and erection of a Blériot, followed by take-off and landing displays in which I took part. Then came speed tests between the monoplane and biplane, reconnaissance exercises including map drawing, photography and plotting positions of troops. Grahame-White demonstrated the possibilities of bombing first by dropping small bags over the side of his machine onto a marked target area, and then he dropped a 100lb 'bomb' in the form of a sandbag from his Farman at a height of 200ft. The machine remained steady as the weight left it, but there were doubts as to whether a high explosive bomb in actual fact would not blast the pilot from his machine before he had chance to get away. For the reconnoitring tests, troops hid over a stretch of land between the aerodrome and St Albans, some 13 miles away, and during these tests their positions and artillery strength were reported. Hamel did a special exercise as an air messenger and delivered an official document to Farnborough. At the same time a messenger left Hendon by land, but Hamel arrived first. Poor Drexel had an embarrassing moment during the demonstrations. He was about to leave for Farnborough when he suddenly shot across the ground in trying to take off. Realising something was radically wrong, he pushed the control column forward to keep the machine down but it had the reverse effect and shot into the air. He proceeded to go up and down in the weirdest possible manner until he finally stalled and crashed. He could have been killed. Fortunately he was none the worse for his misadventure, but the meeting was intended to demonstrate aeroplanes to their best advantage, and show off the skills of pilots in front of the high-ranking Government officials, not this sort of thing. We discovered afterwards that the elevator wires had been crossed on his machine and what he said to his mechanic was nobody's business.

    I had the great pleasure of meeting Cody for the first time. I'd heard so much about him but our paths had never crossed, and what a pleasure it was! I took to him immediately, but unfortunately our meeting was not under the happiest of circumstances since we stood by our machines in the machine park waiting our turn while the main displays were on, and we waited a long time. I'd flown a bit, Cody had not, he'd arrived later than me, but when would we be asked to fly? We hadn't come to watch. I approached the officials, asking when we can be of further assistance. They would let us know. Cody was about fifty, twice my age, well built, a healthy-looking man of 15 stone or so born in America, but the British press confused him constantly with Colonel Cody, known as Buffalo Bill. 'I'm Samuel Cody. He is William Cody.' Cody's plane was a huge pusher with a large elevator in front, known as the *Cathedral* because an earlier machine had its wings in a *katahedral line*, a French term meaning sloping downwards and its name had stuck by popular use. His wife was the first woman in Britain to take a flight in an aeroplane. 'She's the only woman I'll take up. The closeness of the passenger seat behind me is a little too intimate for any lady to be seated in but my wife.'

The demonstration came to an end. Not until many of the people were leaving were we invited to fly. It didn't matter so much to me, but Cody, he was furious, he resented this slight by Grahame-White and took it to heart as he had a great deal of flying experience, more than Grahame-White, having been employed by the Government, and had already displayed military uses of kites for reconnaissance at the beginning of the century and he had built the famous *British Army Aeroplane No 1*. He complained, 'Grahame-White intended this to be a one-man show with himself as the star turn. It's a personal insult to us, having us standing impressively by our machines and then ignoring us.'

I agreed, so did A.V! Grahame-White's name was mud for the time being where Cody was concerned, but our bad treatment did not go unnoticed by *The Aero*:

'The new Roe biplane, which Pixton flew from Brooklands the same morning, distinguished itself by doing a very quick get-off in the demonstration of starting, but after that it was relegated to a corner till the evening when most of the notabilities had departed. S.F. Cody who arrived from Brooklands in magnificent style, had to be content to look on till the evening. However, when he and Pixton did fly, their superiority in speed over the ordinary type of biplane was most marked.'

After a circuit or two Cody had left, disgusted with the whole performance, but we were not the only two who suffered. Barber's machines, the Valkyries, similar to the Farman, had white-coated officials standing in front of them to ensure they would not fly. It would have been better if we hadn't been invited at all. Numerous protests were made and, ironically, Barber presented four of his Valkyries to the War Office, two for the Navy, two for the Army, which became Service Training Machines. And a few days after the demonstrations, I received an invitation from Arthur du Cros to the Houses of Parliament. 'It would afford Mr Lee and myself very great pleasure if you could dine with us at the House of Commons on Tuesday next at 8 o'clock to meet a few Members who are interested in Aviation.'

Grahame-White continued his publicity campaign by flights around coastal towns with his pupils. Painted on his machine was *'Wake Up England'*. This famous slogan was first delivered in a speech at the beginning of the century by King George, then the Prince of Wales, on returning from a world tour when he was appealing for settlers for unexplored countries. Grahame-White, on the other hand, was hoping to arouse the interest of local councils in aviation. Overall, he did a great service for aviation. The Parliamentary Demonstration was a triumph, Hendon was put on the map, the Government was impressed, and the War Office issued a statement to the effect that they would be arranging a contest in the near future to determine what aeroplanes would be suitable for military purposes.

The month of May was hectic. Not only had there been the race to Brighton, the flying display at Haywards Heath, the Parliamentary Demonstration, but the first pilots' strike in history. It happened at Brooklands and I was caught up in the middle of it. It should have never happened but we were led by Mrs Hewlett, and as it would have been rather impolite to argue with a lady, we had to support her. What else could we men do?

The Brooklands Racing Club had notified us that, in addition to the prizes given at each race meeting, 5 per cent of the gross takings of the day would be equally divided between those of us who flew more than fifteen minutes. Mrs Hewlett thought we were not getting a fair deal. *'It's not enough. We want more.'* A letter was drawn up demanding 25 per cent of the gross takings, sixteen strikers signed it, and the *Winning Post* wrote, '*The aero village was seething with suppressed wrath and verdant bile*'. We stirred it up, stating that although Brooklands was built for the motorcar it was, after all, the aeroplane that attracts the crowds and figures prove this. Before the flying, attendance of 5,000 was considered good. Nowadays we drew crowds of 10,000, even 15,000. Where else could one watch a day's motor racing and flying at the same time? Furthermore, those of us who rent sheds at £100 a year were paying for the use of the land to store our machines, to construct machines and to carry out experimental flying. The original contract did not state that we came under the jurisdiction of the Automobile Club to the extent we were now subjected, which includes exhibition flying and an ever-increasing flow of passengers demanding flights. When spectators are milling around our hangars, we are unable to undertake any experimental flying. That was our story...

Major Lindsay Lloyd was taken by surprise. We'd got him really worried and he announced to the press that our letter was the first indication he had that anything was wrong... the club was offering £760 in prizes this season. We refused to fly on Wednesday, the next race day, he refused to negotiate until we withdrew our letter. We did withdraw it, saying that we will fly for the minimum time required to claim the day's prizes. On the race day only Gilmour and myself went up. He flew for seventeen minutes and I did twenty-one minutes, then we refused to let people even look at the machines and locked them all up in the sheds, but the Major had a card up his sleeve.

While we obstinately refused to bring out the machines, we suddenly saw the familiar outline of the *Cathedral*, 'Cody!' The Major had commissioned him to fly and, much to the delight of the crowd, he did a few circuits then gave an exhibition of hedge-hopping, flying as low as possible over obstacles, hangars, bushes, trees, anything in his line of flight, an extremely dangerous type of flying at which he excelled. Cody prepared to make a landing. It was thought that we would turn nasty, and various people, many local navvies, closed in on us and I was surprised to recognise one of them was the proprietor of the Coffee Inn where I stayed. Apparently the Major, prepared for trouble, had hired them to protect Cody. Cody brought his plane to rest near the Blue Bird Café where we were all waiting and his bodyguards surrounded him and his machine, but he broke through. Grinning broadly, he came towards us with the worried bodyguards at his heels and we received him with open arms. He came into the café with us for tea and a chat.

Major Lindsay Lloyd could only be astonished; the hired guard could only be disappointed at having been somewhat deprived, they'd been expecting trouble and might have been looking forward to a spot of violence. Cody clearly knew why he'd been called in and thought it amusing and since he was so popular no one bore dear Cody the slightest malice for strike-breaking. As he left to return to his ground at Laffan's Plain, 15 miles away, he wished us the luck in our fight against establishment.

All was quiet for a while, then we heard Mr Reynolds announce that Cody had flown twenty-one minutes and twenty-nine seconds, twelve seconds longer than I'd flown. None of us imagined that Cody's flight would be counted or that he was even eligible for the Brooklands prizes, but there it was. I wasn't going to have that! Friend or no friend, Cody was taking my £30 prize by twelve seconds and I was set on regaining it, so I unlocked the hangar and took the biplane out and flew for another eighteen minutes. I had to fly at least fifteen minutes, no flight under that counted. In the end my total flying time came to thirty-nine minutes. I had the prize. I pushed the machine back into the shed, locked it up, and no further aeroplane was seen for the rest of the day. Eventually management gave in and everything was settled amicably. We were to receive a fair percentage of the gate money, and the flying prizes were raised so that first prize stood at £50, thanks to Mrs Hewlett. The Automobile Club was not entirely the loser for Brooklands got plenty of free publicity as most dailies covered the rebellion which had been amusing as well as being unique.

Latham, whom Blériot had beaten in the first Channel crossing, paid us a few visits during May and June. He was still flying Antoinettes, and one day raced his Antoinette overhead against Gordon Watney's 60-horsepower Mercedes speeding full out on the Brooklands track below. Then he had a close shave with us. While flying with a friend he switched off the engine to hear a remark and had difficulty switching it on again and just managed to save the situation as he passed over the motor track, but not before touching the telegraph wires at the edge of the field. We ran out to him as he landed, the wing was barely hanging on. *He's brought home an insulator!* Not in the least worried by his narrow escape, he was soon flying again but Major Lindsay Lloyd immediately ordered that the telegraph wires were lowered for safety so that they were just above the ground. Not long afterwards, Latham had another unfortunate experience at Brooklands. He faltered on taking off and crashed on top of the Martin and Handasyde hangar. *'He's done what no other man has done'*, for throughout the history of crashes at Brooklands, no one had ever landed on a hangar. The tail of his graceful Antoinette was sticking up in the air and the engine broke through the roof and was visible from inside the shed.

Although Roe hadn't received many orders for all the time he'd worked on his triplanes, there were signs that the little Avro biplane would attract buyers. The *Westminster Gazette* wrote about it:

'Mr Roe's latest biplane, known as the Avro, in Mr Pixton's hands, is regarded as one of the best-wind punchers extant. It seems to do in the air under similar circumstances what the good old sailing ship does in troubled waters, and that is the highest compliment that one can pay to Mr Roe's ingenuity as designer.'

The first man to buy a triplane from Roe was Captain Windham who'd collected his order just about the time I arrived at Brooklands. Roe had also built a Farman-type machine for a Bolton man and a Curtiss-type biplane for E.W. Wakefield of Kendal who was going to convert it into a hydroplane for experiments on Lake Windermere. He took one of our best pupils, Stanley Adams, with him as his pilot shortly after he'd qualified. Also ready for collection was a new Avro biplane ordered by Commander Schwann who, like Mr Wakefield, was converting it into a hydroplane by fixing floats onto it. Both of these men were pioneers in this field for no one had raised a machine from water. It was a new idea.

The Avro School was prospering. Crashes had come down to a normal level, much to the sorrow of the Brooklands colony who no longer downed tools to watch the Avro School at play. We continued to attract fine men and had secured another machine, a second-hand Farman which belonged to a woman who'd crashed it into a flagpole, and with the extra machine tuition was made easier. Pupils at this period included Kemp, Raynham, Beatty, Sippe and Stanley Adams, also an Indian pupil, Venkata Subba Setti. Then there was Blacker. Blacker looked very distinguished with a monocle he constantly wore and during his tuition he became most impatient as he felt that he had to wait too long before he was due for a lesson and demanded his fee of £50 back. Roe declined to give it to him and Blacker gave poor A.V. a black eye. This was too bad. A.V. was a most inoffensive man and Blacker was so very much bigger. Blacker became disliked after this, his type of attitude was not in keeping with the pleasant atmosphere of Brooklands, and he left to finish his training with Bristol's at Salisbury Plain.

The incident could have been worse for not long afterwards a man was shot dead by a pupil over the very same situation. It happened at Hendon. A Swiss pupil learning to fly at the Blériot School also thought that his tuition was not rapid enough and wanted his fees back. As he did not succeed in getting them refunded, he unexpectedly drew a gun and fired on Monsieur Chereau, the manager, and Petitpierre, works manager. Petitpierre was killed. The pupil then attempted to take his own life by turning the gun on himself. He lay on the ground wounded, drew a razor, gashed his throat and died shortly afterwards, a most unpleasant and unhappy incident in the history of aviation.

Parke had left us and had finished his tuition at Bristol's Brooklands School. He had the idea of setting up as a pilot to take the risks of testing experimental machines and put this ad in *Flight* in May, 'To Inventors. Why break your aeroplane yourself, when we do it for you, Bois Caine Unlimited.' It had no takers.

I decided to leave A.V. I knew I was flying well and had achieved something of a reputation at Brooklands, and thought it time I was earning more. A.V. was not in a position to pay me more than £2 a week as he was still struggling, not for recognition, he'd got that, but for business. I wrote to H.V. Roe in Manchester telling him of my intention, pointing out that I was compelled to consider my future. He replied suggesting that I might become a freelance pilot offering my services to any firm requiring them and flying for his brother as and when needed, but I did not want this. In seeking a new position I approached Captain Wood, whom I knew very well, with the view of joining Vickers, the armament firm who were preparing to branch out into aviation and were intending to test their machines at Brooklands. Captain Wood had been previously attached to Bristol, and was in fact one of their first pupils. He lived close to Brooklands. Like Bristol, Vickers had money to back its plans. Captain Wood had been appointed their technical advisor and he was very pleased to hear I sought a position with them, but wanted me to wait until they became established in their new undertaking. When I said I could not wait he became annoyed, even angry. They had plans to open a Vickers Flying School, but I was not persuaded, it could take until the end of the year and I could not wait that long.

Some days later I contacted Bristol and was offered a handsome salary of £250 with an attractive annual increase of £50. This was much more than the £104 I was receiving, and more than pleased me. I was free to start whenever it was convenient. I told A.V. of my plans, thanking him from the bottom of my heart for the wonderful opportunity he provided which enabled me to enter the flying world at a time it was virtually impossible to find a position and for the experience I'd gained while being with him and how I had enjoyed flying the biplane. I left at the beginning of June 1911, a year after my arrival. Sippe, willing to wait for his licence and for extra machines to arrive at the overcrowded Avro School, was to take over as A.V's Manager, and my work as Avro Test Pilot and School Instructor became the responsibility of Kemp and Raynham, two very capable pilots, both of whom I had taught and had just passed their tests, Kemp, licence No 80, and Raynham No 85.

It was with the greatest reluctance that I severed my close connections with A.V. Roe.

CHAPTER SIX

With Bristol at Brooklands

Bristol's history was not as old as Avros, but they started in 1910 in a big way with the considerable financial backing of £25,000, the first company to enter aviation with ample capital. It was run on very efficient lines and had become the largest manufacturer of aeroplanes in England, and behind it all was Sir George White, a very far-seeing man. Although he'd started from small beginnings, Sir George, the most powerful man in Bristol, owning much of the town, was first known for his work with the Bristol Tramway Company and in organising the transport system of the city. He bought up horse-driven tramways all over the country, introduced the electrification of them, improved the bus services, and then introduced the high-powered Charron taxis making Bristol renowned the world over for its excellent taxicab service. Sir George saw a vast future before the aeroplane, but when he offered their services exclusively to the Government in 1910 he was advised to develop privately as the Government was not ready then to invest in aviation. He started constructing machines at his tramway depot in Filton, first building the French Zodiac under licence – not a successful project – then their own Bristol Box Kites. The immediate running of the firm was left to his son, Stanley White, and with the exception of main races, Bristol did not go in for competitions. Their chief concern was marketing machines.

I was still at Brooklands in my new job, having been appointed Flying Instructor to their Brooklands School. For the first few months my work consisted almost entirely in teaching pupils to fly on the Bristol Box Kites which were basically improved Farman-type pushers. They were perfect examples of faultless workmanship, wrong only in design being similar to the Farman pusher. I took one of them up. I found an entirely different technique was required to handle the Box Kite. Roe's machines had the control column directly in front of the pilot while these had them at the side and when one wing dropped I had to be really ham-fisted, tugging violently at the control three or four times to get the machine back on even keel. Landing was even stranger. Instead of pushing the column back, I had to push it forward so what had become instinct to me had to be forgotten each time I took off or came in to land. I didn't like the machine especially, and certainly favoured the tractor planes more, but of the Box Kites the Bristol was the best, although they did not match many other machines for speed. They were fitted with the French 60-horsepower Gnome rotary engine.

Usually I was out early and often the first in the air, testing the weather conditions before giving instruction on Box Kites. Conditions permitting, I taught about four pupils a day, taking each up for three flights of approximately twenty minutes. All flights were logged. Unlike Roe's planes, the pilot sat in front and on giving pupils their first lesson I took the pilot's seat and the pupil was able to lean forward and familiarise himself with all the controls except the rudder while we were flying. They were good machines for instruction because of this. After flying with me several times, I would then let the pupil get the feel of the machine on the ground and try out the rudder, then I would go up with him seated at the controls while I had the nerve-racking experience of sitting behind as he did the flying, lurching forward only when he was in difficulties. Each pupil had a different temperament which one came to sum up immediately and, thus prepared, the job of teaching was simplified.

The Aero weekly magazine folded as not being a good business proposition while a monthly would be a more profitable concern. C.G. Grey, a joint editor, was given an opportunity to start his own magazine. Victor Sassoon, a great friend of Grey's, offered to finance a new weekly under Grey's editorship. He readily accepted and the magazine, *The Aeroplane* came into being, the first issue coming out on 11 June 1911. C.G. Grey, whom I held in high regard, a man professionally head and shoulders above anyone else in his field, wrote some extremely nice things about my flying, often teasing me, and welcomed anyone calling in to see him – a very kind man to one and all. He still came to Brooklands, of course, and was also going around Hendon and other places to get news. Unfortunately, he never learnt to fly. One aviation journalist did learn to fly, and that was C.C. Turner who wrote for the *Observer, Pall Mall* and *Field*. I knew him, he learnt to fly at the Bristol School at Salisbury Plain and on obtaining his licence he put C.A. after his name. *What is the meaning of C.A?* Certified Aviator! I suppose we could have all done that but, as far as I know, he was the only person who actually did and as a journalist and not a flying man, I expect he felt rather bucked by his achievement, very enterprising of him. He had received licence No 70, April 1911.

Apart from instructing others I was often out flying on my own. Box Kites were very safe in calm conditions, but it was virtually impossible to fly at all in breezy weather as they had little natural stability. I was determined, however, to get some control over the Bristol Box Kite, and to get on terms with it I flew in all kinds of weather. I'd often flown in bad weather in Roe's biplane, but it was in the Box Kite that I developed a reputation for wind flying. The day I was especially noticed was Saturday 17 June 1911, when I flew for seventeen minutes. It was my first flight on the Bristol Box Kite in public, a day when hundreds of overseas visitors for the Coronation of King George V were present. I certainly wrenched at the controls, as I'd discovered on this type of machine, to feel that I was master of it. The flight was very short. I won the £50 Brooklands prize for being the longest in the air but there followed numerous reports on this particular flight of mine:

Clapham Express, 'The large number of Coronation visitors who braved the elements at Brooklands in order to see the aviation which is there making such rapid progress, were well rewarded for their fortitude if only by the magnificent sight which was provided by Mr Pixton. Spectators had almost given up the hope of seeing any flying when Pixton suddenly came out on his Bristol to struggle for the supremacy with what was little short of a hurricane.' *The Standard*, 'The flying on Saturday was some of the most exciting I have ever seen.' *Belfast Weekly Telegraph*, 'Pixton came out on his Bristol biplane to do battle with the strong wind. All eyes were turned towards him. Exclamations of "Oh" were frequent during the seventeen minutes that he remained aloft, a demonstration of pluck which was much appreciated.'

Western Daily Press, 'He made some sweeping dives, causing his machine to come perilously near to the earth, while spectators caught their breath and shut their eyes in anticipatory horror.' (I'd told them that I was not in any real danger.) *The Car*, 'H. Pixton kept the crowd in a state of breathless suspense… His performance was the more remarkable in that this was the first occasion on which he had publicly demonstrated the flying powers of a Bristol aeroplane.' *The Aeroplane*, 'Saturday June 17th Brooklands, nothing doing in the morning and bad wind. The afternoon was very exciting. Brooklands full of Coronation visitors from all corners of the earth, and of all colours, shapes and sizes. Pixton made a hair-raising flight of seventeen minutes in a wind blowing in gusts from thirty to near forty miles an hour… even apathetic motorists in paddock stirred to take an interest, and to neglect bookmakers for a few minutes.'

The Aero, 'Pixton's splendid struggle on his Bristol machine, with strong gusty wind was sensational in extreme, and, though it might be

put down as foolhardy, undoubtedly taught that pilot more than could ever be learned in a dozen flights under dead favourable conditions.'
The Morning Post, 'Happily, after being up seventeen minutes five seconds, he alighted behind the sheds without accident, after the finest exhibition of pilot's skill that has been seen at Weybridge.'

Later, I went up again and added another twenty-two minutes to the Brooklands aggregate time. It all counted and was added to my other times, now at about 225 minutes, for that prize at the end of the season which I still hoped to win. I was well ahead, Raynham was next with just sixty-one minutes to his name. This was about the fourth or fifth race day of the season. The following Saturday, 24 June, I found a willing Bristol mechanic, Briginshaw, to fly with me as my passenger in making up flying time for the Manville competition. Special days were selected for this and this day was one. I flew for twenty-six minutes and later for another five minutes, and although it was June, again it was windy, and more reports came in about my so-called daring flights.

Western Daily Press, 'The English aviator, Mr Pixton, is achieving a reputation as a Wind Flyer, some of his recent feats being really remarkable for their skill and daring…The aviator had scarcely left the ground before the wind sprang up with redoubled force, and it was at once evident to the experienced aviators who were looking on that Pixton was in very great danger. He did not appear to realise it, however, but with the utmost *sangfroid* remained in the air for twenty-six minutes six seconds, a period of the most acute anxiety for all spectators.'

The Autocar, 'The wonderful flight of Mr Pixton, on Saturday afternoon. He flew for twenty-six minutes in a wind velocity of which was seldom less than 30mph. It was one of the most thrilling sights ever seen on the Brooklands flying ground, and for sheer pluck and daring will rank with Latham's exhibition at Blackpool.'

Latham had astonished people at the first flying meeting in England, the 1909 Blackpool Meeting when he flew in wind, an unheard of thing to do. No one flew then, in anything but calm conditions.

The Autocar continued with their story:

'To make such a flight and to see it are all very well, but we do hope Mr Pixton will not take such risks again. One felt one must turn away as a catastrophe seemed so certain, and yet there was a terrible fascination in watching him. Time and again the wind caught him and toyed with him, and still he defied it, righted himself, and went on. We can believe that he enjoyed his battle with the elements, but what must have been the feelings of his passenger, for, incredible though it may seem, he carried a passenger in this terrible flight.'

Briginshaw didn't mind. He wanted to fly with me again, anytime.

'Needless to say, everyone was relieved when he alighted unhurt. Mr Pixton is an especial favourite with the Brooklands flying community. He has been born and bred there, so to speak, and they are proud of him, and may well be so.' *The Aeroplane*, 'Wind and Pixton got up about the same time. A horrible performance, twenty-six minutes six seconds of acute anxiety for most aviators. The machine was blown everyway but the right one, and made appalling dives. Pixton and Briginshaw quite enjoyed the trip. More than others did.'

Various similes were being used to describe the flying such as, *'He was chucked about like a human shuttlecock... buffeted this way and that like a feather... blown about like a piece of paper.'* And one that was often used, *'tossed up and down like a little boat on a choppy sea'*. They would also say, *'flying in lumpy wind conditions'*. I was pleasantly surprised to discover that, as far as one could tell, I'd been the only person to fly on this day in Britain, and Bristol printed a full page advertisement stating this, complete with a photograph of the Box Kite. 'British & Colonial Aeroplane Company Ltd. Going strong. Bristol stands for reliability and strength. This photo shows an all British Bristol Biplane (flown by Mr Pixton) flying in a gusty wind of 30mph at Brooklands June 24th. This was the Only Machine in England Flown on That Day.' I felt rather honoured that they thought fit to advertise their company by using my name after I'd been only two or three weeks in their employment.

I was also put on the front page of the *The Aeroplane*, flying over the sheds at Brooklands. Publicity followed publicity! 'C.H. Pixton. Trained as an engineer, he has now adopted Aviation as a profession. He is held in high esteem by his confreres at Weybridge for his pluck which borders on daredevilry... A brilliant flyer from the first, he is said to be able to fly almost any machine at the first attempt... Mr Pixton, one of the most daring of Bristol pilots... C.H. Pixton, ex-pilot of the Avro biplane and now one of the crack flyers of the Bristols.' I didn't enjoy the bumpy weather and much preferred an easy life and calm days for flying, but most high winds were not troublesome. My dear friend Spottiswoode, who'd flown a bit at Brooklands on an Avis, thought I took too many risks and with his usual friendly greeting, he always said to me, '*The same old Pixton. The same old smile. Still living under false pretences. Still cheating death.*'

One day around this time a man of means arrived at Brooklands looking very affluent in a big, highly-polished silver Rolls Royce. I recognised him as 'Lil Arthur' Johnson, the American black boxing champion, the first black man to win the world heavyweight title which he'd held since 1908. I went over to see whether he wished to fly, to which he replied abruptly, '*Certainly not. I've come to watch. There's nothing difficult about flying.*' However the reports of my wind flying did not put off a Cinematograph Operator from coming up to film, one of the first camera men to do so. It was in the first week of July and he wanted to take aerial pictures for a cinema show, so I flew around with him on the Bristol Box Kite as he filmed the landscape below. The newness of the operation was expressed by the *Bristol Times* and other papers. 'We may consequently look forward to some very novel and interesting pictures to be included in the programme of picture palaces in the near future.'

The first Round Europe Race, organised by France, took place between the 18 June and 7 July 1911. The course covered just over 1,000 miles, starting from Paris and divided into stages over France, Belgium and Holland, the final stages being from Calais to Dover, Shoreham, Hendon, Shoreham, Dover to Calais and back to Paris. Many well-known names were among the thirty-eight starters, but not many got further than the first stage although they could use as many machines as they wanted en route. Amongst the thirty-eight were, de Conneau, Prévost, Garros, Védrines, Weymann, Valentine, Morison, Tabuteau and Tetard.

The terrible toll of three deaths was recorded on the first leg of the race, two of the pilots being burnt to death on crashing. Our man, Tabuteau on his Bristol Box Kite, completed the course, Tetard retired, Valentine got to Hendon with nine others but landed on the return journey at Brooklands with engine trouble, and Morison had crashed very early on. Only nine finished. First was de Conneau, second Garros, third Vidart and fourth Védrines. I was happy to hear that Védrines, a man who'd done exceptionally well in his flying, was placed. When I first met him, he was Loraine's mechanic and he used to wonder whether he would ever possess a machine of his own. He'd led for most of the way.

After the Europe Race there came our turn, the *Daily Mail* sponsored a Round Britain Race scheduled for 22 July to 5 August 1911. It was the biggest aviation event ever held in Britain, covering the whole countryside and a distance of 1,010 miles, about the same mileage as the Round Europe. The start was to be at Brooklands and control points in an anti-clockwise direction at Hendon, Harrogate, Newcastle, Edinburgh, Stirling, Glasgow, Carlisle, Manchester, Bristol, Exeter, Salisbury Plain and Brighton then back to Brooklands, an ambitious undertaking. For the second time Lord Northcliffe was offering the sum of £10,000, the first having been the £10,000 London to Manchester challenge. Our rules were more stringent than those of the European Circuit in which pilots with smashed machines could just get another. Not so for Round Britain. The plane that started had to finish. Not only that, ten parts were to be sealed, five on the engine, five on the plane, and two on each had to be intact at the end of the race, otherwise the contestant would be disqualified. Competitors had to take compulsory rest periods and help could be accepted for repairs but ground time spent on repairs between the controls naturally would count as flying time. Surprises were anticipated, anything could happen.

Lieutenant de Conneau, who was known as Beaumont when flying, and Védrines were entering, so we expected very fierce competition. They were warm favourites having done so excellently in the Round Europe Contest. The Press was really interested. News of the race occupied the attention of thousands of people who were not normally interested in flying, and many who still had never seen an aeroplane would have the opportunity to do so. Days before the start, journalists wrote of the event which was being known as *The Great Race*. However, since the object of the race was to discover the most reliable form of machine, it was also being called, *The Reliability Race*. Such headlines appeared, *The World's Greatest Air Race... Airmen of all Nations... Start on 22 July... A contest of brain skill, rivalry and courage... French and English rivalry... Monoplanes v Biplanes... Can our champions beat Beaumont and Védrines... Can England win?*

The French were exceedingly interested. Their press was covering the race in full, interpreters were to be on hand and crews filming as it progressed. Official information was to be displayed at the *Daily Mail*'s Manchester offices in Deansgate and at their Paris bureau on the Boulevard des Capucines, where model aeroplanes would be moved on a large map as news came in. Preparations along the course began in a big way in Britain. Places chosen as control points were very busy. Harrogate prepared their landing spot on the Stray, the common on the edge of town which they marked with a large white circle 8ft thick. Manchester had trees felled and grass cut at Trafford Park and the ground marked with a large cross, 100yds by 50yds and 20ft wide. Was it really so big? The Bristol control, forty-eight acres at Filton, was preparing to accommodate 40,000 spectators, trams would run day and night into Filton, and hotels and boarding houses around the controls were preparing for a busy time, Police, Scouts, the GPO were all in readiness to cope with the great occasion.

Apart from the major prize of £10,000 several special awards were to be presented, £250 from Sir George White, 125 guineas from British Petroleum, a cup presented by Mr Ogden of Harrogate, also a 50 guinea tea-service for the fastest flights between Hendon and Harrogate, £50 from the Northumberland and Durham Aero Club for the fastest to Newcastle, £100 from Excelda Handkerchief Company for the first to land at Trafford Park. Harry Preston would present a £100 cup on behalf of the hotels of the resort to the first British aviator to reach Brighton, and 100 guineas from the Perrier Table Water Company to be divided between the first Frenchman and first Englishman to complete the course. Probably there were more, too. In all, the Aero Club received about thirty-five entries but several days before the race there were only thirty listed, some having dropped out. Of the thirty entries, eighteen were British, seven French, one American, one British-Argentine, one Swiss, one Austrian and one Dutch. Each paid a fee of £100, the principal engine being used was a Gnome and well over half the machines had them. A scoring booklet was printed in which were details of the pilots and their planes, a map of the circuit, their numbers and nationalities, and short write-ups.

22 July - 5 August 1911

First Round Britain Race

	Pilot	**Nat**	**Plane**	**Comments**
1	BEAUMONT	FR	Blériot monoplane	Lt de Conneau, a fine flyer.
2	ASTLEY	GB	Birdling monoplane	Of considerable experience.
3	MOULINAIS	FR	Morane-Borel monoplane	Recently injured, not likely to start.
4	FENWICK	GB	Handley Page monoplane	A pilot of promise.
5	PORTE	GB	Deperdussin monoplane	A British pilot using a French plane.
6	KEMP	GB	Avro biplane	Took his certificate on a similar plane.
7	PATERSON	GB	Grahame-White biplane	Is likely to do well.
8	MORISON	GB	BRISTOL MONOPLANE	A very fine all round flyer.

9	VEDRINES	FR	Morane-Borel monoplane	Sprang suddenly into notice.
10	RADLEY	GB	Antoinette monoplane	A skilful monoplane pilot.
11	BLANCHET	FR	Breguet biplane	A plane made mainly of steel.
12	CAMMELL	GB	Bériot monoplane	A very promising cross country flyer.
13	AUDEMARS	SW	Blériot monoplane	A very popular man, a fine flyer.
14	VALENTINE	GB	Deperdussin monoplane	Is likely to do well.
15	GILMOUR	GB	BRISTOL BIPLANE	A flyer of exceptional merit and dash.
16	ENGLAND	AR	BRISTOL BIPLANE	Sound hard working cross country flyer.
17	PIZEY	GB	BRISTOL BIPLANE	An absolutely fearless flyer.
18	PRIER	FR	BRISTOL MONOPLANE	An excellent flyer.
19	PIXTON	GB	BRISTOL BIPLANE	Born with the instinct of flying.
20	CODY	GB	Cody biplane	Designer, and pilot of his own machines.
21	TABUTEAU	FR	BRISTOL BIPLANE	A magnificent biplane flyer.
22	JENKINS	GB	Blackburn monoplane	Has made some good flights.
23	MONTALENT	FR	Breguet biplane	Another French pilot driving a Breguet.
24	HAMEL	GB	Blériot monoplane	A very fine monoplane flyer.
25	REYNOLDS	GB	Howard Wright biplane	Of the Royal Engineers.
26	LORAINE	GB	Nieuport monoplane	The well known actor, a fine flyer.
27	HUCKS	GB	Blackburn monoplane	Recently took his certificate at Filey.
28	WEYMANN	US	Nieuport monoplane	A modest man, first class flyer.
29	WIJNMAALEN	DT	Deperdussin monoplane	An excellent flyer, Dutch.
30	BIER	AU	Etrich monoplane	Daring Austrian flyer.

A race of seven nationalities, machines of three nations, fifteen British, fourteen French, one Austrian. Order of starting was drawn by lots and I would leave just before Cody. In the company's efforts to place their machines in the fore, Bristol had entered seven men, Prier and Morison on monoplanes, Tabuteau, Gilmour, England, Pizey and myself on biplanes, and we had a splendid organisation of supply wagons and mechanics who would be standing by in readiness at the controls and at other selected spots. My machine was to be Tabuteau's Round Europe biplane, it was still scrawled with signatures from the crowds it met during the European circuit. Tabuteau was in France waiting for another race to start which had been postponed and was hoping to make it in time for the Round Britain and Gordon England was flying for Argentina. As he was born there and both his parents were English, he could adopt either nationality, and since he'd been approached by Argentine representatives in Britain to fly for Argentina, he saw no reason not to accept.

Unfortunately, Gilmour's chances of entering were nil since he'd had his licence suspended by the Aero Club and was not allowed to fly. He'd joined Bristol in February and at the beginning of the month displeased the authorities by swooping low over the Henley Regatta. An Aero Club rule stated that no person should fly unnecessarily over thickly populated districts or areas where a number of people gathered, but he ignored the ruling much to his regret. Gilmour was a lively lad and had displeased local authorities before, landing once at Hampton Court in the middle of a herd of deer, yet I could not help thinking the suspension was unfair as, at the time of the Cambridge and Oxford boat race earlier in the year, five men from Hendon led by Grahame-White, including Gilmour, had flown over the race and must have certainly broken the rules as much as Gilmour had done at Henley. Poor Gilmour was angry. Bristol took his case to court and finally to the Court of Appeal to argue the fact that the Aero Club had no right to suspend a licence which they did not grant, his licence having been issued by the French Aero Club. It was to no avail. The Aero Club was supported by the International Aeronautical Federation so the rules applied to him. Gilmour could not fly and half in fun and half in sorrow, he put a black crepe wreath over his shed at Brooklands to signify his grief at being prohibited from entering the race.

Preparations at Brooklands began long before the race. Huge packing cases were arriving followed by pilots and mechanics, and when unpacked and assembled, Brooklands was filled with an assortment of interesting machines, some being given a fresh coat of paint before their trial flights. Our French compatriots were friendly and curious, wandering into our sheds with smiles and some, without a word of English, showed such interest and willingness that they helped us with our machines as well as attending to their own. We learnt that weeks before any race Beaumont turned teetotal, a non-smoker, and slept as much as possible. I passed him several times sound asleep in the sun beside his Blériot. After a day's work many pilots and their helpers made themselves as comfortable as possible, finding mattresses or quite happily sleeping on straw or sacking, often with ten or more dossing down in one shed, and some set themselves up quite nicely in empty packing cases which served as bedrooms. All mechanics had to be at Brooklands three days before the race unless unavoidable delay could be proven. Harold Perrin and Alec Ogilvie were among the nine members of the Aero Club who commandeered our machines while they sealed the ten parts on each and painted our numbers on the fuselage in large black letters.

Saturday, the day of the race and thousands were arriving. Bus routes had been extended, three extra trains per hour were running from Waterloo to Weybridge at a cheap return fare of two shillings and every imaginable kind of vehicle, charabancs, cars, taxis, motorbikes, cycles and horse carriages, were to be seen. Even the 20 miles to Hendon, our first stop, was filling with people and their vehicles. Every available seat at Brooklands was taken, luncheons were served in a marquee and buffet stands were everywhere to cater for peoples' needs as they waited. Weymann, the American who'd just won the Gordon Bennett race at Eastchurch, was astounded on seeing the crowds gather, *'I've flown in many great races in Europe, but I've never seen anything like this.'* And it was so hot, one of the hottest days I'd ever known in England, a record 92 degrees in the shade. People fanned themselves and ladies put up sunshades. A hot, windless day was no good for flying as the sun had the effect of creating currents and unexpected eddies which the general public would not suppose exist. One of C.G. Grey's men made an attempt at expressing the heat of the day. *'Brooklands had become one huge saucepan, and the air was simply boiling out of it, just as a pot boils.'* The weather forecast was 'light south-westerly winds and much sunshine'.

Just before the start Beaumont suddenly discovered he'd lost his cap and when we went over to see what the fuss was about we heard him say, 'It is missing, I always fly in it. I am fondly attached to it, I am very unhappy about my loss. It is not valuable it cost me 2.50 franc, anyone bringing it to me, I will give 100 franc.' The fact that it was missing and the finder would be rewarded was announced over the speaker, but poor Beaumont was never to see his favourite cap again. The race was to start at 3pm.

Unfortunately Bristol had been badly hit. Out of our team of seven, four were unable to compete. Gilmour prohibited, Prier damaged his plane in the morning, Tabuteau was still in France and Morison had got a cinder in his eye and was laid up; that left only Pizey, England and me. Of the other contestants the Avro man, Kemp, was also out of it. On making a trial trip at about 1pm he struck an air pocket, a wing collapsed, and he came down sideways, crashing into the ground. Roe's biplane was a wreck, Kemp was unharmed. I was very surprised having flown the machine, but apparently Roe had added wing extensions and one had come adrift. Once again the little biplane had lost an opportunity of proving itself. There were four other contestants out of the run: Moulinais was injured in France, Fenwick's machine was not ready in time, Radley had crashed his Antoinette and Loraine his Nieuport. That made in all twenty-one potential starters out of thirty. The four Bristols, the one Avro and four others were out of it, then before the intended starting time, an announcement was made. There would be one hour's postponement. Instead of starting at 3pm, we were to leave at 4pm when the air would be slightly cooler after the mid-afternoon's heat.

Line up, fifteen minutes beforehand. At the stroke of 4pm, timekeeper Ebblewhite was ready to send us off at four-minute intervals on our first stage to Hendon. First away was Beaumont. The crowd cheered and waved everything wave-able... next Astley, then Porte but he got caught in an eddy and crashed, fortunately not over the crowds. Paterson was next, then Vedrines, who flew into the eddy which had brought Porte down... then Blanchet left sluggishly just missing Captain Wood's house and surrounding trees before gaining height, the trouble being he'd not tuned his engine sufficiently... Cammell, Audemars, Valentine, then England but he was unable to get away because of engine failure and it seemed as though only two of us would be left representing Bristol. Pizey got off nicely but had a little trouble in controlling his machine as he headed for Hendon, then I shot off with an appalling noise from the newly fitted but effective 60-horsepower Renault. Behind me, the most popular man of the race, Cody. The start was described by the *Daily Mail*:

> 'Perfect weather from dawn to sunset. A vast gathering of spectators, far greater than has ever been at Brooklands before. A mile and a half of motorcars round one side of the track. The hill thronged with people. The sheds enclosure crowded. A total attendance estimated at between 40-50,000, and the best exhibition of the marvellous art of flying that has ever been given in this country. In no country, indeed, has anything better been seen than the prompt departure at the fall of the flag of so many of the world's aeroplanes piloted by the most famous airmen of the day.

I wonder how many of the spectators realised what a miracle this was. I doubt whether anyone who has not followed flying closely can quite realise it. When one recollects how uncertain aeroplanes were less than two years ago, how capricious their motors, how haphazard their control, it is like a dream to see them line up and one after the other soar into the air at the word Go.'

The *Daily Mail* reporters at Hendon took up the story as we flew in, having this to say as Cody and I came in:

'A Bristol is next in view and sails low over the sheds and crowds. Mr Pixton's tanned (and pleasant) face appears in the car, and we are level with the foreigners. A monster comes next, rolling in the breeze and gliding down over us with a vast and shining steel propeller slowly turning. It is a weapon to make an executioner's mouth water. So deadly a thing has no right, we feel, so near us. "Cody in his Cathedral", cries a voice, and there is a scamper of photographers.'

I got my card marked by timekeeper Reynolds, and as I was walking towards the enclosures I heard the rattle of a car behind me. The next thing I knew was being lifted completely off my feet by a massive hand locked under my chin. I fell against the car and was affectionately kissed, much to my embarrassment, in front of press cameras. It was Cody. This moment was captured and went to press, *The Tatler* heading their pictures, '*The Penalty of Fame*'. Mr Cody fondly and forcibly kisses Mr Pixton at Hendon on his arrival. Mr Grahame-White appears to enjoy the joke.'

Grahame-White, who was in the car with Cody and others, was not competing but had organised everything at Hendon including where we were to stay over the weekend before making a serious start early Monday morning. During the race we would land at the controls, check in and leave as soon as we could, taking our rest periods when we wished. A latecomer from Brooklands was Weymann in his Nieuport who came in at a great speed just after 6pm and he told me of his little misfortune on starting off. He was nicely in the air when his map unrolled and blew all over the place and he had to return to get it fixed and make a new start. The damning thing was that his flight was the quickest to Hendon, taking him only fourteen minutes, whereas Vedrines took nineteen but Weymann had to be timed from his first take off. Jenkins, Reynolds and Hucks were still at Brooklands waiting to start when the sun was lower, as they said there was 'no grip' in the air. It was a well-known fact that heat affected the efficiency of an engine. Many certainly did have a difficult start because of the immense heat of the day and out of 21 starters only 17 finally arrived at Hendon. Jenkins, in his attempt to reach Hendon, had crashed and Cammell, an Army contestant, came in last around 8pm having been forced down between Brooklands and Hendon because of a broken valve spring. Perhaps 50,000 *had* watched us at Brooklands. Some doubled the figure and others doubled even that for the total number who'd seen us this particular Saturday.

On Sunday I set to work on the ailerons of my Bristol as they'd required considerable effort to work them, more than usual, that is. I managed to reduce the leverage by half and took the machine for a flight to test them out. They functioned much better. On coming in to land, I exercised the greatest care as I was very much aware of the Hendon ground and its bumps, so much rougher than the Brooklands ground, and I didn't want a broken chassis at this early stage of the race. What an extraordinary enthusiasm was shown at 4am, early Monday morning, as we lined up to start on the second stage. Dawn was breaking. Colindale Avenue leading to the ground was thronged with private cars and the fields were covered with people. Men, women and children had camped out, lit night fires, made coffee, snatched some sleep, prepared their food, just to be there when we started. There was a buzz of excitement, and groups of French men and women sang the *Marseillaise* and shouted from time to time, *'Vive la France'*.

The order of leaving Hendon was fixed by the time taken on Saturday from Brooklands and all went well. Beaumont shot off swiftly. Vedrines should have left first though, but he followed Beaumont, then Hamel went, Valentine, Audemars, then I left, seven minutes after Beaumont. My Bristol got away beautifully, and just as I was thinking it was a first class start, the engine spluttered, so I turned back to get the trouble put right. With a fresh set of spark plugs, I made a second start and was on my way over the long track across country to Harrogate, the third control. A haze hung over the land and it looked as though it could be foggy, quite a contrast to Saturday's heat and blue skies. After two hours flying, I'd covered 90 miles and reached Melton Mowbray. It was about 6am, and I made a landing on the polo ground. By arrangements, this was the first filling station for the Bristol competitors only.

The local paper, the *Melton Mowbray Times*, took an interest in the happenings of the day in their area:

'The next sensational local happening was the arrival on the polo ground at 6am of C. Howard Pixton, who came down in his Bristol biplane for petrol having left Hendon at 4.10am. Hundreds of people who had hitherto remained outside the ground made a frantic rush through the gate, those in charge being helpless to cope with the throng, and unable to collect the sixpences due for admission... many hundreds of admirers were soon swarming around his machine, and they watched with interest the work of replenishing the engine with petrol.'

A cup of coffee and biscuits were put in my hand by a Bristol representative. Within fourteen minutes I was off again, having topped up with four gallons of petrol. The misty conditions seemed to be clearing as the day progressed, but I was still running into dense patches over towns. Fortunately, my compass was reliable this time, and the wrist altimeter I'd recently bought was proving to be most effective.

I was flying at a fairly high altitude and keeping a good course, but suddenly over Spofforth, roughly 4 miles short of Harrogate, the engine cut clean out without warning. I had to find a landing ground, and find one quickly. I saw a cricket field ahead of me, and began gliding towards it but just as the wheels were about to touch the grass, the engine started up again, so I pushed the column back and climbed, thinking I could reach Harrogate before investigating the trouble. A few minutes later, the engine cut out again. There was something radically wrong. I knew under these circumstances I should not risk a turn, but I could see no suitable landing place ahead of me and therefore decided to return to the cricket pitch. The machine stalled. Down I went, missing the Spofforth pavilion by inches, and plunged into the turf. Both of my legs were trapped and could be broken. I imagined the worst. A strut had pierced my thigh. I pulled it out. It had entered the flesh an inch or two. The leg bled a little. I groped around for a handkerchief to stop the bleeding as blood trickled down my leg. I struggled free. My legs were not broken. I glanced at my watch. It was about 7.55am. 'Are you all right, sir?' I turned and faced a policeman. A crowd had silently assembled. He was Constable Bettley.

Apart from the punctured thigh, my middle finger had quite a severe gash and my back was sore. I stared at the crumbled Bristol, ruined! The trouble proved to be the simple matter of a shortage of petrol. *The tank was dry!* I had eighteen gallons at Hendon, I'd used nine gallons over the 91 miles to Melton Mowbray, which was about halfway to Harrogate, so I should have had ample as I'd topped up at Melton Mowbray with four gallons. The tank must have leaked on the way... the engine had restarted as I was going in to land as the tilt of the machine sending enough petrol into the carburettor to make it fire again. Of course I should have realised this and landed there and then, instead of trying to get to Harrogate. However, I was out of the race and, with my wounds neatly bandaged, I was taken to Harrogate. An enormous crowd had gathered there, and as we came into view an official rushed over to the car. I gave him a brief account of what had happened for the record, and was put to bed at the Prince of Wales Hotel overlooking the Stray. I had to have two stitches in my punctured thigh, and then the news poured in.

Soon my bed was covered with newspapers in which were several reports of the Spofforth incident:

Leeds Mercury, 'There was great excitement at Spofforth and the neighbouring Barkston Ash villages when, after three airmen had successfully passed on their way to Harrogate, Mr C. Howard Pixton in his biplane came to grief. He first appeared coming very slowly over Wetherby Workhouse and at a small altitude, for his number could be distinctly seen as he crossed Spofforth Hill, and went to the Deighton side of Stockeld Lodge. He was travelling so slowly that it was not surprising when he was seen to drop near Spofforth. He rose again, but once more collapsed.'

Yorkshire Herald, 'The aviator found himself in difficulty as he was directly over the village, so he called out to the people below to direct him to a safe field.'

I had to smile. This was pure journalese or a misinformed account but I had had a narrow escape. It was the worst crash of my flying career and it could have been my last. The *Onlooker* had a picture of my crashed plane captioned, *Why Pixton got no further than Spofforth!* It had been an exciting time for the public. Headlines of the day's flights appeared everywhere. '*Great Morning Flight... Fog and Wind... How Early Risers caught a glimpse of the race... Thousands of Mansfield people rise early... Keen interest in Nottingham... Magnificent flights... Vast crowds all the way...*'

Spectators between Hendon and Harrogate had made their own deductions as to whether our machines would pass over their areas. Hundreds went to high places and stood waiting and watching for hours. Many saw only one machine whilst others were lucky to see three or four pass their way, as some of us had deviated from the direct course, flying more to the east or more to the west. Many had not turned up for work, shopkeepers had notices in their windows, '*These premises will be shut because of the race'*, and reporters were kept busy dashing from one place to another in the fastest cars available. Telephone exchanges beat all records as people rang friends telling them to hasten over as a machine had landed in their area. It was reported that while one machine was grounded for repairs a little boy was seen twisting a seal off a vital part of the machine, and one pilot was reputed to have said, 'So thick was the fog that I quite thought I was about 1,000ft up when my machine touched something with a thud, and I found myself on the ground in a field.'

From the hotel I had a good view of the control but nothing much was happening now. Védrines, Beaumont and Valentine had long left Harrogate and were heading for Edinburgh. Cody had landed at Martin's School in search of the control and had to be directed to it, and he stayed at Harrogate to repair a leaking tank. Hamel had arrived too, collapsed, but later continued. Pizey, our last hope for Bristol's, reached Melton Mowbray but had to retire. Bristol had gone to so much trouble, having men positioned all around the course to look after us, and except for those at Melton Mowbray, they had had nothing to do, I being the last of the Bristol men on the scene. By the end of Monday, the state of the contest was drastic. Nearly everyone was out of the race, and the countryside to Harrogate was littered with machines. Only the three leaders were flying well, free from problems.

22 July - 5 August 1911

First Round Britain Race 2nd Day

Bristol's 7 Men

	Position	Miles	Pilot	Comments
1	Edinburgh	363	VEDRINES	Still mobile.
2	Edinburgh	363	BEAUMONT	Still mobile.
3	Edinburgh	363	VALENTINE	Still mobile.
4	Newcastle	270	HAMEL	Still mobile.
5	Harrogate	202	CODY	Still mobile.
6	**Spofforth**	**196**	**PIXTON**	**Crashed**
7	Wetherby	196	MONTALENT	Had lost his way, smashed propeller.
8	Leeds	184	WEYMANN	Had lost his way, damaged chassis.
9	Wakefield	175	CAMMELL	Burst cylinder and overturned machine.
10	**Melton Mowbray**	106	**PIZEY**	**Retired**
11	Irthlingborough	72	ASTLEY	Lost his way, hoping to continue.
12	Streatley	48	BLANCHET	Engine trouble, broken wing.
13	Burton	46	HUCKS	Damaged machine.

14	Codicot	35	BIER	Wrecked machine.
15	Hendon	20	REYNOLDS	Delayed start.
16	Hendon	20	AUDEMARS	Retired.
17	Hendon	20	PATERSON	Retired.
18	Leaving Brooklands	0	PORTE	Crashed.
19	Leaving Brooklands	0	JENKINS	Crashed.
20	**Brooklands**	**0**	**ENGLAND**	**Engine trouble prevented start.**
21	Brooklands	0	WIJNMAALEN	Engine trouble prevented start.
22	Non starter	0	KEMP	Crashed before race.
23	**Non starter**	**0**	**PRIER**	**Damaged machine.**
24	Non starter	0	LORAINE	Crashed before race.
25	Non starter	0	RADLEY	Crashed before race.
26	Non starter	0	FENWICK	Machine not ready.
27	**Non starter**	**0**	**TABUTEAU**	**Absent.**
28	Non starter	0	MONLINAIS	Injured in France.
29	**Non starter**	**0**	**MORISON**	**Eye injury.**
30	**Non starter**	**0**	**GILMOUR**	**Prohibited.**

I'd not done so badly to get to Spofforth, but how I regretted not having done better. Several minor inaccuracies were reported, but this was the type of occasion when rumours travelled like wildfire, silenced only by official reports. In fact word got around that Valentine had been killed, entirely unfounded, and I too. As someone put it, *'And they had even made away with Poor Pixton'*. I left Harrogate before the race finished and went to Bristol to give exhibition flights there arranged by the *Daily Mirror*. The local paper reported on it. 'That Mr Pixton still feels the effects of his accident was noticeable from the fact he walked stiffly. He was the reverse when up, however, and treated the spectators to one of the finest performances that has been given in Bristol.'

The race continued as a contest between four men only, Védrines, Beaumont, Valentine and Cody. Védrines and Beaumont were clearly the leaders, both making good pace and flying magnificently. I was not at Brooklands when, on Wednesday, Beaumont flew in on his Deperdussin, the winner. He was received by Lord Northcliffe having flown the 1,010 mile course on his Blériot in four days, flying time 22 hours 28 minutes, his average speed, 45 miles an hour. He'd flown the British course at a much faster pace than the European Circuit of about the same distance which had taken him over

58 hours to complete. Not long afterwards Védrines came in 2nd on his fast Morane-Borel monoplane.

In two years £21,000 of the *Daily Mail*'s prizes had gone into French banks – Channel crossing £1,000, London to Manchester £10,000 and now round Britain £10,000 – a staggering blow to the British aeroplane industry!

While Beaumont and Védrines were safely at Brooklands, Cody was only as far as Newcastle where he was giving exhibition flights, and Valentine had arrived at Carlisle. By Friday Cody was at Edinburgh and Valentine was still at Carlisle delayed by mechanical troubles, but by the following Tuesday Cody had left Carlisle as Valentine departed from Gloucester. Next day, a week after the winner's arrival, Cody had reached Bristol and Valentine was well ahead at Exeter. The following day as Cody came down at Weston-Super-Mare, Valentine had reached Salisbury Plain and was nearly home. Friday saw Cody at Salisbury Plain and Valentine in trouble at Horsham, 20 miles short of Brooklands. He came into Brooklands by car to pick up mechanics and returned to Horsham, then he flew his Deperdussin triumphantly into Brooklands in third place, the first British pilot to get around the course. I saw his machine a few days later. It was in such a shabby state that it looked quite unsafe to fly. Next day, Saturday, 5 August and the last day the race was open, Cody came in arriving at Brooklands so early that no one expected him, no official was there to greet him. He was the only British competitor on an all-British aeroplane who'd succeeded in completing the course, and it had taken him a fortnight. Poor Cody was fagged out, but his first cheerful words were, '*I guess that wind's enough to shake Pixton up!*' He'd suffered very heavy damages and did all his own repairs, yet despite everything all his seals were intact.

First Round Britain Race Results		
Position	**Pilot**	**Arrival Date**
1st	BEAUMONT	26 July
2nd	VEDRINES	26 July
3rd	VALENTINE	4 August
4th	CODY	5 August

And so two Frenchmen had won, and taking away with them Lord Northcliffe's £10,000 prize money! That was that, the end of the most noteworthy competitive event of early British aviation, a race which received huge press coverage in Britain, France and America. A great race, a race free from fatalities, and a race during which one elderly man declared, '*Now I have seen the greatest invention of the 20th century in full flight, I can die happy*'. Although Britain did not win, it had not been without its rewards, as one high-class magazine put it, 'Flight achieves its first real success as a popular spectacle.'

My crashed plane, a complete write-off, was brought back to Brooklands by Bristol mechanics, but not before many had come from far and wide to see it at Spofforth. This was to be expected. The fascination of an aeroplane drew people like bees to a honey pot, especially in country areas. Resulting from long-distance flying across country, such as the distances that had been covered by the Round Europe and the Round Britain Race, it was suggested by the *Federation Aeronautique Internationale* that all railway stations should have the names of the station written on their roofs for the benefit of pilots, and that large buildings have names of towns on them.

The race was not quite over when I returned to Brooklands, and I'd just got back when, much to my horror, I witnessed the first fatality there. In spite of the tremendous amount of flying done at Brooklands, no one had been seriously injured although we all knew that sooner or later we were certain to have a death. It had to come and I, or anyone else, could have been the victim. It happened on 1 August 1911, the man who died was Gerald Napier, a young lad of nineteen whom I'd just taught to fly. He obtained Licence No 104 and had bought a machine of his own, a Bristol single-seater biplane designed by Challenger and Tabuteau. Napier wanted to fly it immediately. I volunteered to try the machine out thoroughly for him in order to compare the feel of it with the one to which he'd been accustomed, since it was much faster and smaller, but I wasn't prepared to try it on this particular day as the weather was not very good for flying. He wanted to try it out for himself.

After making a few unsteady trial flights, I was rather surprised to see that he was about to go up with his friend, Laurie, who was sitting astride the petrol tank hanging on to whatever he could. I went over to him, pointing out it was a single-seater and not made to accommodate a passenger, that he was taking a terrible risk by putting undue strain on the structure. But no, he would not listen and waved gaily taking off with his friend. I had no power to stop him, then it came. After doing a circuit, he stalled as he turned too sharply and from a low height, side-slipped to earth. I was first to the wreckage followed closely by John Alcock and others. Laurie had been thrown clear, stunned, but okay. Napier was underneath the wreckage. We lifted the structure off him, a splinter of ashwood from the undercarriage had entered his mouth and pierced his head. He must have died instantly.

When the inquest was held at Brooklands a verdict of accidental death was recorded. As a witness and as Napier's instructor, I told the coroner what had happened, saying the deceased, a pupil of mine, purchased the machine ten days ago and had made only three short flights immediately before the accident. This first death in our circle upset each one of us. No flying was done for the rest of the day and out of respect the Brooklands flag was flown at half-mast. It was the fifth death to occur involving British pilots, others being Rolls, Grace, Benson at Hendon, and Smith in St Petersburg. This unhappy death of Napier gave rise to superstition among some of the men by the reason that his machine had been put in Gilmour's shed on delivery on which the black crepe wreath still hung since the suspension of Gilmour's licence. Some thought this was significant. It was taken down and promptly burnt.

I'd already taught a number of people to fly with Bristol, but one of my more distinguished pupils was Captain Brooke-Popham who joined us in June and had received Licence No 108 in July. He flew very well and, as he thanked me for his tuition, presented me with a watch which was simply inscribed, 'C.H.P from R.B.P. 1911.' Another distinguished pupil was Brigadier General David Henderson, an older man of about fifty. He'd been in the Army all his life and had come to me in August under the assumed name of Henry Davidson, hiding his identity in order to avoid preferential treatment. This was typical of him. He was one of the most considerate men I have ever had the pleasure of meeting. As he lived in Byfleet not far from the track, there were no problems of travel, and during his tuition I met his son and took him for several flights. They were exceptionally fond of each other. Henderson, a born pilot with 'beautiful hands', learnt to fly after only a week's training and, in doing so, became the first British general to hold a flying licence. It was No 118 and the magazines did not miss reporting this occasion.

> 'Brigadier General David Henderson, Chief Staff Officer to Sir John French, stands out prominently as the first officer of such high rank to take up flying practically, and his passing for the brevet in one week is quite a remarkable performance, for which Mr C.H. Pixton, his instructor, deserves his share of praise and congratulation.'

As most pilots still wore their caps back to front to prevent them from blowing off when flying, the brigadier had turned to me saying, *'Now I'm entitled to wear my cap the wrong way round'*. In the same month as Brigadier Henderson received his licence, Mrs Hewlett, the strike leader, got hers, No 122, during August and became the first woman to qualify in Britain. She'd used a Hewlett-Blondeau-built Farman, and a few months afterwards having taught her son on it, he gained his Licence No 156. The Hewlett-Blondeau

School at Brooklands had taught ten pupils and was free from accidents and major repairs, which said a lot for the Farman-type machine for teaching. The Bristol Box Kites were good too. They were so similar to the Farman that it was often difficult to tell them apart.

I hadn't landed in the sewage beds since flying the Roe triplanes, but when I took up Gordon England's Round Britain plane, which had refused to start for the race, it let me down. The machine would have been nice to fly with a little more horsepower, but it was too late as on a turn I found myself heading into the sewage beds. No great damage done, but *Flight* picked up the story, '*Pixton thought he would just clear the sewage farm. The engine thought otherwise*'. My engines have been accused of many things. When I'd recently come down near the railway station, someone wrote, '*The engine took a rest*', and at Spofforth, '*The engine shut up shop*', but this was the first time that my engine had 'thought otherwise'. The sewage farm incident brought out a further humorous touch in *The Aeroplane*, 'Rumour had it that Pixton was looking for his pet *remous*.' A *remous* was a little eddy. They continued, 'The *remous* is really not so bad when you get to know it. Treated with patience and kind words, it becomes quite docile, and will make itself useful in a lot of ways. It is getting so much attached to Pixton that the other night it followed him nearly half way back to the sheds. But the statement that it was found next morning curled up under the left wing of his Bristol, fast asleep, is an exaggeration.'

I had been looking for a good second hand car, as I could afford one now with my increased salary and money won during race days at Brooklands, so when I heard a man at Weybridge had one to sell, I went to see him at the local garage where he kept it. The garage owner said I could take a run in it and tell him the next day whether I liked it or not. That was fair enough. My intention was to try it out on the Brooklands track but I'd just driven about half a mile when it broke down, so I walked back to the garage and told the owner what had happened. He seemed taken aback. I told him I was not interested in buying it. Next day, I received a letter from the garage saying that I'd taken delivery of the car and would I pay the money for ownership. Also, it pointed out, 'You will be charged five shillings a week until it is removed from the garage.' Here, I thought, was someone trying to con me into buying a useless vehicle, and as I did not want it, nor wanted to pay for it, I went to the Police. It turned out that the man in question had actually forged the proprietor's signature and had stolen the garage's headed paper to write to me. The man had no business relationship with the proprietor except by that of payment for storing the car. The Sergeant who looked into the matter was very surprised. Mr X was a very well-known figure in Weybridge. I don't know what happened, nor cared to know, but no further demands were made.

Soon afterwards I bought an Argyll which ran beautifully and proved to be a most reliable car, although a bit slow. One of my first passengers was a dog, an Irish terrier called Whisky. It was a funny little thing that attached itself to the Bristol wagon bringing mechanics to work from Weybridge. Whisky did not belong to any of the men but was owned by the licensee of the public house where some of them stayed. He would literally take possession of the wagon, snapping and snarling at one as he guarded his adopted property. I used to annoy him as I did Gilmour's wolf-dog Seti, but in this case by just placing the tip of my forefinger onto the wagon. This would immediately set Whisky off. If I held my finger but a quarter of an inch away, he was all right. Not until my finger actually touched the wagon did Whisky bark and rush madly up and down. I supposed I shouldn't have teased him so, but I always made a fuss of him on the ground. In fact, we were really the best of friends as it was so easy to get him off the wagon with a friendly call, and when I was not busy I took him for trips not only in the Argyll, but in the Bristol Box Kite. He loved nothing better than going for joyrides, and when flying he never fidgeted nor caused me any anxiety whatsoever. I always had half an eye on him though, just in case, for I would have been terribly upset had anything happened to him.

It was during August that Bristol decided to transfer me to their flying ground at Larkhill, Salisbury Plain, 60 miles away. I didn't like the idea in the least for, not only did I consider Brooklands my home, but this change meant parting from my friends. Brooklands was unique, a great place, its fascination gripped me as it did everyone who worked there, and a more dreary, desolate spot than Larkhill I could not imagine. I was to continue giving flying instruction, and replacing me at Brooklands were Pizey and Fleming, 'Little Appy' and 'Big Appy.' Happy they may be, I was not when on 31 August the time came for me to leave. I put my personal belongings into the Argyll and set off for the Plain not quite in the best of spirits. Both *Flight* and *The Aeroplane* made pleasant remarks about my move, 'Mr Pixton, who had hitherto been practically a fixture at Brooklands, has left to go as Instructor to the Salisbury Plain School, and he will be very much missed by his many friends at Brooklands, where, in spite of his splendid flying, he never lost his peculiarly modest manner and, because of it, won himself hosts of friends... Mr Pixton, *The Shining Light of Brooklands*, has gone to the Salisbury School of the Bristol Company... Pixton left for Salisbury, much to everyone's regret...' all very nice, but nothing could cheer me, an air of gloom descended upon me.

I did not want to go!

CHAPTER SEVEN

With Bristol at Salisbury Plain

Larkhill was bleak, its bare undulating land with hardly a tree on it looked an uninspiring place to teach flying. The nearest village, Amesbury, was 3 miles away, and Salisbury about 8 miles. Once again I was lucky to find accommodation close to my work in one of the houses nearest to the flying ground called 'Crossroads', next door to the Stonehenge Inn, which was kept by the wife of one of our mechanics. I settled in and took up my duties as Flying Instructor. Bristol leasing the ground from the War Office, had set up a group of five hangars. Not far from them was another group occupied by a section of the Air Battalion and staffed by Captain Burke, Captain Massey, Lieutenant Barrington-Kennett, and Lieutenants Cammell and Reynolds both of whom were competitors in the Round Britain Race. Bristol supplied the Air Battalion with two or three Box Kites for their air manoeuvres and target practice on the Plain, and was the first company to get an order from the Government. Stonehenge lay two miles to the south-west and was visible from Larkhill. So as not to interfere with the historical tradition of the monument, a space had been left between the two groups of hangars allowing the rising sun to strike the altar at a certain time of the year. This space was known as the Sun Gap.

Besides being the largest company in Britain manufacturing machines, Bristol's schools were the busiest and taught about one third of all the pilots in the country. The Larkhill school was especially busy. Several officers came to us from the two Army camps close by, Bulford and Tidworth, so we were never short of pupils. As many as ten, would be waiting each day for tuition. Larkhill was not only Bristol's school ground, but also their testing-base for new machines which were transported from Filton. Mechanics unloaded and assembled them, we tested them.

I worked closely with an Australian called Busteed who'd recently been appointed School Instructor. We would go up on trial flights in our Box Kites to see if the weather was suitable for flying, then start the men off on a morning's training. When outdoor work was impossible, we gave what instruction we could on machines on the ground. I knew most of the Bristol staff before coming to Larkhill, including the four Frenchmen, Tetard, Tabuteau, Jullerot and Prier. Tetard and Tabuteau had been with the company for a considerable time, but around this period they'd returned to France, Jullerot (Jolly Rot to us), who had worked at the Mercedes factory in Germany, had been engaged by Bristol when they first set up at the end of 1910 to test Box Kites, and when the schools were opened was put in charge of their running. Prier was the company's latest designer. He had worked with Blériot on his early monoplanes and during September 1910, came to England to take charge of the Blériot Flying School at Hendon, then in April 1911, established a name for himself on becoming the first man to fly non-stop from London to Paris, and had set up a new cross-country record of 250 miles. He joined Bristol just a few weeks before me and, backed by excellent experience of the Blériot machines, started work on monoplanes ignoring the Box Kite principles which had established Bristol in business. Although I was accustomed to the Bristol Box Kite, I did not feel attached to them and was looking forward to the day when the Bristol-Prier would be in wider use.

Busteed, England, Gilmour, Valentine and Morison were all Bristol pilots, and all well-known at Brooklands, but England had done less flying on taking up designing, also I knew Harry Delacombe, Bristol's Flying Manager who'd joined the company in a temporary capacity. Others on the staff included Challenger and Pizey. Challenger, formerly of the Bristol Tramway staff, was now the firm's Aero Constructional Engineer at Filton with the Bristol-Challenger 'T' type Box Kite to his name, quite a successful machine, only six were made, four of which had been entered for the Round Britain Race. Pizey, a Somerset man also formerly of the Tramway staff, was Assistant Engineer and Superintendent of Construction at the Filton works, but when he learnt to fly he became the Chief Instructor at Larkhill before replacing me at Brooklands.

Captain Dickson had joined the company shortly after his accident in Milan to become the firm's Technical Advisor, supervising design at Filton. A great friend of Sir George, although he looked perfectly well, he suffered great physical discomfort from internal injuries since his terrible midair collision and had to have a stomach pump which his valet attended to, clearing out his stomach every few hours. I'd met him on several occasions and not long after my arrival at Larkhill I had the pleasure of taking him for a flight, his first since the dreadful accident. He joked about it, *'Just to see whether I've lost my nerve or not'*. Throughout the flight, he remained cheerful even though he must have been grieving knowing that he might never control a machine again. It was tragic. He really loved flying. *The*

Aeroplane was particularly interested in this flight, and he told them how he felt about it:

> 'I'm very pleased with my flight and have great confidence in Pixton as a pilot. I feel scarcely strong enough to control a machine for any length of time, but I certainly hope to do a little flying, now I know how I feel in the air again.'

Another man whom I knew well at the Plain was Cockburn. He had a hangar near us and, offering his services free to the military authorities, had given instruction to several officers on his Farman, but he left Larkhill for Eastchurch to help in the training of Naval officers there. I had not been long at Salisbury Plain when Lieutenant Cammell of the Air Battalion was killed. Lieutenant Cammell crashed at Hendon 17 September 1911 in taking delivery of a Valkyrie for Farnborough. He handled it as though it were his Blériot to which he was accustomed, flying swiftly and taking tight turns, but the machine did not stand up to this type of treatment. It was helpful towards a long life never to strain a machine, never to stall and never to turn back when the motor fails after take off, three lessons that flying had taught me, which I passed on to my pupils.

Larkhill was quiet, but Bristol always had ample transport for their men to Larkhill and Brooklands. Their wagons took three in the front and six at the back, not counting the hangers-on, and the men used them in their free time whenever they wanted to travel to the nearby towns. A popular place always packed with airmen, was The George at Amesbury run by Mrs Whistler, which had become almost as famous as the Blue Bird Café of Brooklands, and autographs were always sought after by the locals when pilots were recognised. Well, very quickly my opinion of Larkhill changed. There were not the crowds I'd been accustomed to at Brooklands, but I came to realise that the area was exceptional for flying, and found that I actually enjoyed the peace and quiet. When chatting one day with C.G. Grey, I told him this and he reported my conversation with him, '*He is quite in love with Salisbury Plain as a Flying Ground, and believes it to be the finest possible training ground for cross-country flyers.*'

This year, 1911, was the Pioneering Year of the Waterplane and E.W. Wakefield, who had the Curtiss-type machine built for him by Roe, played a great part in advancing the waterplane. His interest began when, troubled by the number of injuries sustained through crashes on land and possible death of pilots, he thought that if a machine rose from water, and flew over water, the degree of injuries would be considerably lessened, so he had started up the Lakes Flying Company at Windermere. Although swamped with letters of complaints that he was spoiling the charms of Windermere by his experiments, he succeeded in raising the Roe-built plane fitted with floats, from the water. There were doubts as to who was the first in Britain to leave from water, but both Wakefield and Commander Schwann are popularly credited with being the first but Schwann, who also used an Avro biplane, was under the impression that Wakefield was first. In his own words Wakefield credits neither himself nor Schwann:

'The *Water Bird* made its first successful flight from the water on 25 November 1911, which was eight or nine days after my friend, Commander Schwann, rose from the water at Barrow on his Avro, but nearly five months after Mr Oscar Gnosspelius, who is designer and engineer to the Lakes Flying Company, first left the water in a machine of his own construction, altogether different from the present machine, but this first machine has not yet done anything more than long hops.'

Long hops – just as Roe had done as his first attempts at flying. However, persistent as ever that they were the first to fly in Britain, Avro now advertised, '*First off land. First off*'.

Dennistoune Burney of the Royal Navy, a pupil of mine and son of Admiral Sir Cecil Burney, was a man of many ideas who was closely linked with Bristol as he endeavoured to connect the aeroplane with the sea. They said of him, '*He was born with salt in his eyes*'. He worked on machines with folding wings for easy deck transport and hydrofoils for take-off in rough seas. His ideas were backed by the Admiralty and patented jointly by Bristol and Burney himself, and were developed in a secret design office, 'X Department', with the help of Frank Barnwell. Frank was new at Bristol and was one of the Barnwell brothers who had designed very early experimental machines in Scotland which were not very successful. Neither brother had applied for a flying licence, though Frank intended to do so soon. I had several adventures with Burney, my first when the company asked me to follow up one of his hunches concerning a submarine which had gone down near Portsmouth off Hayling Island. Burney believed we would be able to locate its exact position for the Navy from the air. It was to be the first time that an aeroplane would be used for a naval matter of this nature. So, Burney and I left Amesbury on a Box Kite for Hayling Island. It was a windy day and we made slow progress battling against a strong headwind, then stopping at Durley for lunch and battling on, we eventually landed outside the Royal

Hotel on the seafront, and created a sensation. Crowds collected wondering why we'd come, and during the following days more came, many on bicycles, especially to see us.

Farnell Thurston, Bristol's Sales Representative, Briginshaw and Mackintosh, our mechanics, later met up with Burney and me, but we were unable to do much as the awful weather continued. We camped the machine on the shore close to a tea place, pegging it down securely and pulling up bathing machines to act as a wind screen. Day after day we hung around waiting for the weather to clear but it continued to blow a gale. However, we made one adjustment to the machine which helped to pass the time away. In the event of coming down in the sea during the search, we'd fitted two airbags under the wings but I found they created too much drag, so we took them off and replaced one centrally under the seat which proved better and would provide ample buoyancy. Finally we managed to survey the area and flew out to Nab Lighthouse, but the water was so murky with the sea having stirred up the sand that there was no hope of spotting the submarine. We decided to give it a few more days, then abandoned our task and flew home. I supplied photos whenever I could and brief accounts of anything interesting occurring to the flight magazines, and as this looked an interesting venture we had pictures taken showing the Bristol outside the hotel, by the bathing huts, the crowds gathered around, and the machine showing the new position of the float. They were gladly received and printed but the Navy took no further interest, and the submarine remained where it was indefinitely.

Herbert Oxley, our hearty Yorkshire man, our friend and my first pupil, was killed in a crash at the beginning of December, the eighth Briton to meet his death in a flying accident. He was twenty-five and unmarried. He died when both wings of his Blackburn monoplane tore off while flying with a pupil as he flattened out from a steep dive over the cliffs of Filey. The wrecked plane fell on the beach below. Oxley had left us, switching to Gordon England's School where he got his licence, No 78, on the Hanriot monoplane and had joined Blackburn's as their test pilot. Had he lived, he would have been a famous pilot. Crashes were due to three causes, engine failure, faulty piloting and faulty machines. It was not uncommon that structural weaknesses caused accidents, and was more common for manufacturers to blame pilots and the Blackburn people put the blame onto Oxley instead of their weak machine, accusing him of stunting. Although aircraft were well-built generally, one danger was the breaking up of machines midair, and one could not stress enough how inadvisable it was to strain them. This is what Oxley had done, as had Napier and Cammell. I knew Robert Blackburn well, and he'd flown with Oxley many times, and was most upset by his death. Oxley and his pupil were the first to die on one of their machines, and in expressing his regret, he wrote to *Flight* describing the incident, calculating that on flattening out the total stress on the wings was eighteen tons.

> 'With reference to Mr Oxley's terrible accident, I am enclosing a letter from Mr Hunt, and after going through all the facts myself and considering the description of the accident, there is undoubtedly only one conclusion to be drawn from it. It appears that Oxley was very fond of making sudden and steep descents at a very acute angle, and had lately done a great many of these from moderate heights with several of the aerodrome men as passengers. You will quite understand that when making one of these steep descents and suddenly flattening the machine out, the strain must have been enormous. Although the manoeuvre had been successfully accomplished many times by Oxley, yet there is a height from which velocity at the end of the dive would be so great that no machine could possibly stand it. These descents had no doubt given him such self-confidence that neither he nor others seem to have realised the danger. The facts are that in this case he did a sudden dive over the town of Filey from a height of 600ft. Considering the position from which he started to descend and the position on the sands where he finally fell, the angle could not have been any other than very acute, as the machine was wrecked only about 100ft from the cliff's side. At this point the cliffs and houses at the top are about 350ft high.
>
> 'It seems hard to blame him in such a case, but there is no doubt he was rather fond of giving the onlookers a sensation, and I have been told on many occasions how people have thought that he was rushing down to his death.'

Sir George Cayley had once said, *'before flying becomes safe, at least a hundred necks will be broken'.* Still several people in Britain opposed flying and one, writing to a journal signing himself *Terra Firma*, asked, 'Is Aviation immoral? Is it moral, by the bait of considerable sums of money and prizes to lure young men, fiery with intrepidity, to a death, if not certain, at least probable? Is it not cruel to tempt foolhardy men to their deaths by these gilded baits? Are we to demand gladiatorial sacrifice from our aviators?'

Flying was considered a too-dangerous occupation to follow and it was well-known that many men who would have liked to take it up as a career refrained from doing so for the sake of their families. *'Aviators should not marry'*, was often stated. A new man at Brooklands learning to fly with us called himself Fred Merriam, his real name, Frederick Warren. He had changed his name so his relatives did not know he was flying since it would have worried them. An old friend of mine, Sam Hall whom I saw infrequently and who followed my flying with great interest and enthusiasm, would say to me, *'You'll break your crazy neck one of these days, Pixton'*. Then there were strong views about women flying. 'Aviation is man's work... No woman has the right at the present time to engage in Aviation either as a pilot or passenger,' but it would take more than a few words to stop Mrs Hewlett from flying, or indeed pilots from taking lady passengers for flights. Many crashes were still 'safe' though, and I wished I'd been there when our friend, E.V. Sassoon, the millionaire known as Smith who was always crashing the *Big Bat* and never coming to any harm, struck it unlucky once more around this time at Brooklands. *The Aeroplane's* report:

> 'Mr E.V. Smith on *Big Bat Blériot* going quite well around and about and over the sheds, but apparently tried to land 20ft up in the air. Flattened out angle of machine, switched off, pan caked, dived suddenly, buckled a fork, broke the propeller, pirouetted on one wing tip, cut the cloth rather badly, collapsed some more of chassis, and finally wound up in front of the *Blue Bird*, absolutely unhurt after the most prolonged smash on record.'

The year 1911 was coming to an end and it had been a good one for me. My wind flights had paid off not only during the Brooklands Race Days, but for the Manville Contest, and my transfer to Larkhill had not interfered with my flying and prevented me from competing for the prizes to be won at Brooklands. I would leave Salisbury Plain in the morning to make sure of the prizes and would experience what was known as the *Basingstoke Bump* half-way between the two flying fields where the machine would suddenly drop due to the change of air, it was quite a jolt and only felt on Box Kites. I'd heard others talk about it, but now I knew what they meant. So, I was often at Brooklands seeing the old faces and doing the necessary duration flying and returning to the Plain having won money, and when the last day of the season was over, it was obvious I'd taken a substantial amount.

The Manville, the all-British passenger-carrying competition of aggregate flying time, was staggered over eight specific days and had started at Easter and continued into October, roughly one a month. We had to reach a qualifying fifteen-minute flight as we did for the Brooklands flying, to count as anything, also conditions stipulated that the weight of both pilot and passenger should not be less than 20 stone. Ogilvie and Barber had entered, but it eventually became a contest between Cody and me, and at the end of the August bank holiday, a Manville day, Cody nearly caught up with me as I spent much of the day trying to tune up the ENV engine on the Bristol Box Kite. It was giving enough power but after a few minutes running got hot and smoke poured from it. I changed the mixture and took the carburettor to pieces, but still it wouldn't function properly. I'd gone up three or four times with my passenger and could only manage flights of less than fifteen minutes. I was wasting time so decided to give up the Manville flight for the day and put in for the Brooklands prize instead, flying solo, and did much better, taking the first prize easily with Raynham flying against me.

We flew anywhere for the Manville contest if observed by officials. Cody, who usually did his flying on Laffan's Plain, came to Brooklands on this day and did three flights totalling 80 minutes which raised his Manville flying time to 150 minutes against mine of 187 minutes. There was one more Manville day to go, arranged to take place on 4 October 1911 and Cody was determined to put up a fight to take the prize. I had to see he didn't! It was the last motor race meeting of the season at Brooklands. I'd been lucky to find a constant and congenial passenger in Briginshaw, but he was otherwise engaged this day, so Lieutenant Harford, Royal Artillery, who was due to return to India at the beginning of November, volunteered to fly with me. We went up twice and were tossed about considerably, but most of the Manville days had been windy and this day was no exception. It was cold and wet in the bargain, but though the flights were rough Harford told *The Aeroplane*:

> 'The only reason that the flights looked so alarming was that Pixton did not correct the pitching and tossing of the machine until it really needed it. My only complaint was the intense cold. I have now developed such an admiration for the Bristol teaching methods that I will definitely join the Bristol School at Brooklands.'

And he did, working very hard he got his licence, No 152, within the month before leaving for India and a week before Mrs Hewlett's son had got his licence No 156.

After flying with Harford, I had a spell with Captain Richy of the Indian Army who had just got his licence at the Brooklands Bristol School, and eventually made my total time for the Manville this day two hours nine minutes, not bad going for a rough day, and before the day was out I also added 108 minutes to my Brooklands flying time. At 5.30pm the Manville competition closed. Meanwhile, Cody was competing against me at Laffan's Plain, Farnborough, but only added forty minutes to his time… and he had gone to the bother of finding a young boy weighing not much more than four or five stone to match his 16 stone for minimum weight. The poor boy, accompanied by his mother, had waited all day at Cody's shed as Cody was put off by the wind. Flying had seemed out of the question for him until about 4pm, but when he was all ready to go up, his engine wouldn't start. Not until 4.50pm at the last minute, did he fly. He had no idea of the time I'd put in until after the contest when he discovered I was the easy winner.

1911

Easter to October

Mr Manville's Aggregate Passenger Carrying Competition

Date	PIXTON	Results	
6 May	31 minutes	1st PIXTON	5 hours 16 minutes
20 May	49 minutes	2nd CODY	3 hours 10 minutes
24 June	31 minutes		
15 July	76 minutes		
4 October	129 minutes		

1911

Easter to October

Brooklands Motor Club Aggregate Competition

Date	PIXTON	Results	
17 April	87 minutes 32 seconds	1st PIXTON	7 hours 33 minutes
10 May	40 minutes 17 seconds	2nd RAYNHAM	3 hours 44 minutes
5 June	58 minutes 22 seconds	3rd NOEL	1 hour 34 minutes
17 June	39 minutes 15 seconds		
20 July	16 minutes 17 seconds		
7 August	103 minutes 20 seconds		
4 October	108 minutes 40 seconds		

HOWARD PIXTON **Prize Money for 1911**	
Manville Aggregate	£500
Brooklands Aggregate	£150
Race Day Prizes	£269 7s 11p
Total	£919 7s 11p

My total prize money was just short of £1,000, at £919. I had earned more money than anyone else flying in Britain during 1911. I'd flown the Avro biplane at the beginning of the season for the Manville, but had used the Bristol Box Kite since the summer. Nevertheless, Roe was pleased with the results and was reported so, 'A.V. Roe, exceedingly jubilant over Wednesday's results, announced that not only had three Avro pupils carried off first, second and third prizes in the Brooklands Aggregate Flight competitions, but that his wife had presented him with a daughter on the same day.' A lot of publicity ensued with front page coverage:

> 'Remarkable Flying at Brooklands... Mr Pixton's Triple Success... The Outstanding Performance of the Day.' Bristol advertised, 'three fresh triumphs for Bristol aeroplanes! Manville Prize won by a Bristol Biplane! Brooklands Aggregate won by a Bristol Biplane! Day's aggregate, won by a Bristol Biplane!' The Engine Company ENV also advertised referring to the last day of the season. 'All-British Aviation Motors. Manville Prize, £500. Brooklands Racing Season Aggregate, £150. Brooklands Duration Prize, £30. All won by C. Howard Pixton, using a 60 hp ENV.'

The ENVs were nice engines with copper-covered water jackets, and had been very popular but were fading from the scene around this time.

So ended the 1911 Racing season with my taking the major portion of the prizes offered by the Brooklands Automobile Club, which boiled down to the fact that I'd flown before the public more often than any of my friends. From the flying point of view, however, the summer had been a bad one, very hot and windy, but this was to my advantage as I flew when my rivals would not leave the ground. On 14 December the Royal Aero Club Annual Dinner was held which happened to fall on my birthday and presentations to be made during the occasion were announced:

'Manville Prize £500 to C.H. Pixton. British Empire Michelin Trophy No 1 £500 to S.F. Cody. British Empire Michelin Trophy No 2 £400 to S.F. Cody.'

Ogilvie, Maitland, McClean, Porte, Brabazon, Dunne, Barber, Morison, Radley, Gilmour, Howard Wright, Roe, and other flying men and constructors, MPs and Army personnel, made up the 300 guests.

Cody and I had much to talk about, and during dinner he urged me to join forces with him. I was very fond of him and naturally very flattered. I'm afraid I had to say no and told him that I thought his *Cathedrals* were the wrong design. The Wright brothers and Cody's machines, and a few others, had lifting tails in front unlike the Box Kite which had an elevator in front and a square tail. When a gust of wind hit the front of a lifting tail machine it was raised up and as the gust travelled along, the back was lifted producing a curious pitching movement. 'Tail first machines are not the thing of the day now. They remind me of the Oosely bird which flies tail first to keep the dust out of its eyes,' I told him, he laughed, as he always did whenever I spoke my mind about his *Cathedrals* which had won so many prizes for him, and did not take offence at my reluctance to join him. A remarkable person, and on every occasion we met my affection for him deepened. One could not help liking and admiring him and to everyone he was affectionately known as *Papa Cody*.

It was nearing midnight when the prizes were awarded. I was first on the agenda. Mr Manville, who was President of the Society of Motor Manufacturers and Traders, presented me with the £500 prize, and *The Aeroplane* reported:

'Mr Ernest Manville, after a graceful speech, presented his prize of £500 to Mr Howard Pixton, who, being a modest youth, and at all times a man of few words and doughty deeds, returned thanks in a still voice which did not penetrate to my end of the room, albeit I have known him make himself heard above the roar of a Gnome engine when in his natural element.'

I wasn't addressing the whole room, a personal thank you to Mr Manville was all that was required, and I had no intention of making a speech, unlike Cody, who always enjoyed making speeches. His turn came next, and he was presented with his awards amounting to £900 by Mr Wolff on behalf of the Michelin Tyre Company. About him *The Aeroplane*, reported:

'Cody's utterances in reply were, under the circumstances, a model of self-restraint. He merely remarked that owing to the late hour he would omit some of the things he might have said if he had come earlier in the programme, a neat touch that, for the man of the moment to class himself deliberately with the artists who were booked for purposes of entertainment. He thanked Mr Wolff, referred briefly to the fight for the Manville Prize, said how he appreciated the sportsmanlike way Pixton had fought him for it, shook hands publicly with Pixton, regretted that no one had put up a fight with him for the Michelin Prizes, told us that when Lieutenant Parke smashed the smaller Cody biplane it was not the fault of bad piloting, but merely an accident that he himself or his friend Pixton might have had, warned us all to look out for the next appearance of the Cody machine, and then strode majestically back to his place.'

I'd entered both the Michelin contests which also closed in October, but found I was kept much too busy to seriously compete in either of them. The first contest, open over a period of five years for the longest distance flown each year in a closed circuit on a British-built plane, was in its second year, and Cody won it flying five hours non-stop over a 7-mile circuit clocking up 261 miles. The other stipulated that the fastest around a circuit of 125 miles on a British-built plane over one of four routes at Eastchurch, Brooklands, Hendon or Amesbury, would claim the award. There'd been eight entrants attempting this, but only Cody had flown a complete course and it took him three hours to do so. Cody's winnings for the year were substantial, but it did not escape the notice of the *The Aeroplane* that I'd won more than any other British pilot in Britain.

> 'Pixton is the winner of the biggest prize money of any British Aviator this year, except, of course Mr Grahame-White and Mr Sopwith, who won their money in America, but he remains the same modest, unassuming Pixton of old, and has by no means lost his fondness for Brooklands and its old associations.'

Sopwith, on his return from America at the end of 1911, spoke to *The Aeroplane* about the Wright brothers. No longer with a competitive machine the brothers, who once electrified the world, were dying from the scene. The report was revealing:

> 'Mr T. Sopwith, who had only arrived from America the previous day, gave a most interesting account of his experiences. Evidently aviation is making great strides in the States, but it is being done on American lines which certainly would not work in this country. The Wright

Brothers have adopted a most extraordinary method of disposing of their machines, and seem to have established, if not an actual acknowledgement of their patents, at any rate, a very healthy fear of the consequences of infringing them. Mr Sopwith says that anybody who wants to fly a Wright machine has to purchase it for £1,000 and thereafter has to pay a royalty of $100 (about £20) every time the machine is taken out for public flying, competition or with prospects of profit, i.e. for passenger flights etc. Consequently, if there happens to be a £25 prize up for a small competition such as for a starting competition, and so forth, and there are six competitors, the one who wins gets £5 after paying the Wrights their appearance fee of £20 and the rest lose £20 each, while the Wright brothers make £120. It looks fairly good business for the Wright brothers. However, the Wright machines have got such a name in the States that aviators are apparently quite willing to handle them even on such terms, though their willingness is somewhat aided by the fact that if they dare to fly anything else except a Wright, a law action is promptly taken against them. Those who can afford to defend merely enter a defence, and are apparently left alone, but the poor ones are compelled to knuckle under.'

I understood the brothers tried to pin down Grahame-White, but had been unsuccessful in gaining the sympathy of the courts in a suit against him in respect of the Farman and Blériots he'd used while flying in America. Taking all this and other things I'd heard made up what appeared rather unpleasant history where money and not flying had taken grip of two historic figures, who were doing much to the detriment of American flying.

So 1911 came to an end. Constructors were flying less and specialists were appearing. Flying schools became recognised training centres, 168 pilots had licences in Britain, fees of thousands of pounds for appearances had more or less come to an end, huge prizes becoming a thing of the past having done a great service in promoting British aviation, with Grahame-White, the most successful of all the early British flyers, having made a fortune. Yes, I had had a very good year but, unlike Grahame-White, my flying career began a year too late to become wealthy by it.

The extraordinary of 1910 was now the ordinary of 1911.

CHAPTER EIGHT

With Bristol Overseas

Bristol, with their go-ahead methods and capital doubling each year, were taking the lead in aviation. The company's reputation for flying tuition was secured, we were favourably regarded by the Government having taught many officers to fly and had gained a reputation in other countries as well as Britain as several overseas pupils came to us. One such pupil was Zee Yee Lee and I'd given him some instruction. He was sent by his Government to train with Bristol and later became the Chief Flying Instructor at Nan Yuan Military School. The flight magazines recorded his success on receiving his licence. 'Mr Zee Yee Lee, a Chinese pupil at the Bristol School on Salisbury Plain, has now passed for his certificate, and having done so is the first subject of the Emperor of China to become a Certificated Pilot Aviator.' He'd gained licence No 148 in October 1911.

However, our plans for the future during 1912 were an all-out effort to capture an overseas market for our aeroplanes, and a deciding factor in this direction was the introduction of the Bristol-Prier monoplanes, the first really good machine we had produced which created great enthusiasm when it appeared on its trials at Larkhill. One had been ordered for the Air Battalion. That was encouraging for us, but many people were still disappointed by official attitude. Up to now, the Government had lagged behind in positive attitudes towards aviation, and C.G. Grey expressed impatience about the Government's slowness in the following manner:

'The aerial defence force of Great Britain has received or, rather, is about to receive two important additions. During the past fortnight or so the War Office has ordered two real flying machines, one a Bristol two-seater monoplane of the most recent type, and the other a two-seater Deperdussin. When these are delivered they will bring our fleet of up to date aircraft to a total number of four, and we shall then be within reasonable distance of Roumanian Air Power. At this rate of progress we may, supposing other nations to be standing still, approach the air power of France or Germany in approximately twenty years' time.'

Then *Flight* took up the Naval aspect of flying.

'There are some things they do better in France and aerial development for national defence is one. It is announced that during the current year France is to spend £40,000 on experiments in connection with Naval aeroplanes. Up to the present there is nothing to indicate that we are to be allocated for similar purposes as many pence, while there is an ever growing agitation on foot against privately financed experiments which are being conducted on Lake Windermere. We dare venture to the opinion that were Windermere in France, not only would there be no agitation, but that money, both public and private, would be offered freely for the helping forward of serious and useful experiment. We are credited with being a patriot people, but.... So far as concerns the official attitude towards the Naval aeroplane, we are not inclined to pessimism, because the signs are not altogether wanting that the Admiralty is on the eve of a serious move in the required direction. But the process is very, very slow.'

They were right. There were signs of movement in the Government quarters, for 1912 was the year in which results of sound thinking and solidarity were made evident, the first major move on the part of the Government being the announcement of a forthcoming Military Aeroplane Trials, a competition open to the world to determine the best type of aeroplane for military use. The trials were to take place during August 1912 when submitted machines would undergo vigorous tests and companies producing successful types would be rewarded in prizes amounting to £9,000, also Government orders would be placed with the winning companies. Bristol, who had once wanted to work exclusively for the Government, were naturally very much interested, and had a few months in which to produce the finest machines possible for entry.

Meanwhile, it was overseas for me. Not yet were other British aero firms established enough to enter the foreign market in an efficient and profitable manner, and in this aspect Bristol were pioneers in this field. As I'd flown the Bristol-Prier on several occasions during their tests, I was quite accustomed to handling them and was called upon by the company to play a part in promoting them. Farnell Thurston, the Bristol's representative, briefed me on my immediate missions. He said that in our bid to secure orders from European countries and persuade their Governments to build Bristols under licence, the company wanted the Prier monoplane demonstrated before them, and that I had been picked as the right man for this job. Bristol favoured the French Gnome rotary engines which were kept cool by rotation of the cylinders with radial fins, the old water-cooled method having been abandoned, and this type was being fitted to the Prier machines and proving good. Thurston said he would accompany me on most of the missions and at least two mechanics, with Frank Coles as the one in charge, would be with us wherever we went. With the help of agents in each country, all the arrangements would be done beforehand, including hotel, boat and train reservations, with the machines ready and waiting for us on arrival. All I had to do was to fly.

The prospect of travelling again came as a very pleasant surprise and, with this new type of life in front of me and increased salary, the future looked bright. By taking on these trips, I became among the first to show off a British machine abroad, and first no doubt, a British tractor plane. We were to go to Spain, but before that I was to relieve Jimmy Valentine in Paris. He was promoting the Bristol-Prier while the 3rd International Paris Aero Show was taking place, by flying around Paris in a machine similar to the one on display at the show. Jimmy was a fine pilot. I'd known him in the early days at Brooklands when he was learning to fly with Macfie and he'd beaten me to it, securing his Licence No 47 just a few days before I got mine. He was doing a good job and had stirred up considerable enthusiasm with the Bristol-Prier, so much so that the Bristol stand had become the centre of interest, where people realised that the machine they'd been craning their necks to look at was British and not French.

We stayed at the Hotel d'Albe, half-way down the Champs Élysées, and as soon as possible I went to the show at the Grand Palais where our machine was the only one among forty-three exhibits representing Britain. I made a straight line for the stand and there stood our Prier looking very elegant. It had been especially done up for the show and how it shone! Attached to a highly polished dashboard and surrounds were impressive-looking binoculars, vacuum flask, map case, a small writing tablet, a compass and airspeed indicator, all of the latest style and of the finest quality then, setting off the machine handsomely, were the seats beautifully covered in pigskin. It was actually the sensation of the show, cheap too selling at £950, less than our Box Kite which went for £1,100. Even the French, who led the field in aircraft design, acknowledged it as a superior machine.

Although the Prier attracted attention, the French exhibits, mainly heavy and strong-looking monoplanes, were very much more advanced than the average British plane and included the latest Nieuport, Deperdussin, Morane-

Saulnier, REP (Robert Esnault Pelterie), Breguet, Borel, Zodiac, Blériot and the Farman monoplane. The Blériot and Deperdussin Popular sold at £350 and the highest priced machine was a three-seater Deperdussin at £1,800. One cleverly designed biplane was the creation of Coanda, a Roumanian, and I was especially interested in seeing it having learnt he was joining the Bristol Company. It was an interesting machine driven by a turbine engine. A uniformed commissionaire stood at our stand, an Englishman whose fine physique was a credit to Britain, but apparently the Parisians had never seen a commissionaire in such a smart uniform for, poor fellow, they embarrassed him terribly by standing in front of him, staring steadily, even touching and poking him as though they'd not made up their minds whether he was real or not. I met up with Gordon England, he'd been responsible for bringing over the commissionaire and the cases of catalogues for the Bristol stand. Stanley White, Delacombe and Jullerot were also present to see things went smoothly. After spending a few hours there, I left, having thoroughly enjoyed going around and having a first-class view of what Britain was up against in the market.

As there was such a demand for Jimmy's flying, he'd been invited to give exhibition flights at Saint Cyr, a ground with military dirigible sheds and a few hangars on it, and the place where Santos-Dumont developed his famous but not very successful Demoiselle monoplane, nicknamed the *Infuriated Grasshopper*. I went to Saint Cyr to help out. I was leaving for the flying ground in a little car which the Bristol agent in Paris had arranged for my use when, going through Place de la Concorde, a Paris taxi-driver ran straight into me. No one was hurt, but the collision attracted a big crowd. Unfortunately I could not speak French and when Gendarmes arrived I found I was being arrested and became highly indignant as I felt they should have directed the weight of law against the taxi-driver and not against me. Ignoring my protests, I was marched off to a nearby Police Station and charged with being a foreigner driving without a licence. No one seemed bothered about my car which was just left in the middle of the road for anyone to run into. When I realised they were not going to let me go, I did not know quite what to do, but I indicated I wanted to make a phone call. After much trouble, I finally got through to the proprietor of my hotel. I told him of my plight and he immediately rushed to the station to my aid. After a brief but rapidly spoken conversation with the Gendarmes, there was a complete change of attitude. Everyone was smiling and shaking hands with me, and I was free to go. It was all very puzzling, and afterwards I asked my rescuer however he managed it. '*Ah, Monsieur Pixton, we French, we love aviators.*'

Busteed, Licence No 94, was in Paris too, as we were to continue together to Spain afterwards. All three of us, Busteed, Valentine and I, explored the city in the evenings, visiting nightclubs and sometimes attending concerts as Busteed was exceptionally keen on music. He would get extremely annoyed during a concert when people coughed or sneezed in the middle of a performance. One evening they were determined to get me drunk and took me out with that intention, paying for me wherever we went. I liked that very much, but while I remained sober, poor Valentine and Busteed got as drunk as lords and I had to lead them back safely to the hotel. They couldn't believe it...! One day Jimmy and I went to the Eiffel Tower. As the lifts were only working up to the first stage, we decided to walk to the second gallery where we noticed a woman apparently, like ourselves, appreciating the view. We stayed for a short time at the second gallery, then walked down again, but as we were half way to the first stage, something hurtled passed us and neither of us realised what it was. On reaching the bottom, the battered body of the woman whom we'd left on the floor above, lay in front of us and as we looked horrified at the sight, hardly crediting what had happened, officials covered her up with a coat in what seemed almost a casual manner. We were told that suicides from the tower were quite frequent and were dealt with calm efficiency, but Jimmy and I left the place feeling unpleasantly revolted.

Jimmy stayed in Paris. Busteed and I made our way to Spain with Farnell Thurston to demonstrate the Prier monoplane and the Bristol Box Kite, our destination the Cuatro Vientos Airfield near Madrid, 3,000ft or so above sea level. The name meant Four Winds, not quite an ideal spot for a flying ground if it lived up to its name. We travelled on a *wagon-lit* through France, and when morning came I discovered how difficult it was to shave with a cut-throat razor as we jogged along. While in Madrid I had my first real experience of continental driving. We'd hired a car to take us to the airfield and came to the conclusion the chauffeur must be completely mad. He drove hell for leather hooting loudly over the extremely bad roads full of potholes, having as little regard for his own life as he did for ours, and we could hardly believe it when he lifted a pedestrian off the road with the front of the vehicle. What was more surprising was witnessing the pedestrian humbly raising his cap and apologising for being in the way! We got to the airfield in one piece, and there flew before the King of Spain, Alfonso XIII, Busteed in the Box Kite and I in the Prier. The king personally took a keen interest in the machines, discussing them in detail with us, and spoke extensively with Thurston. The Spanish Government possessed hardly any machines, about two I believe.

After several days demonstrating the Bristols, taking up officers and members of the government, we made arrangements to depart and left the two machines with the military authorities. It was decided, however, that Busteed should stay on to take control of the Spanish School of Military Aviation with complete charge of training officers on the Bristol machines. Not long afterwards we heard that Busteed had taken the Avia Cup for flying from the Celibes Statue to the Royal Palace and back, an offer across Madrid which had been unclaimed in Spain for over a year. It had taken him about half an hour. Our trip had been a success and orders were placed for several Priers and Box Kites.

From Spain we continued to Germany. Our agent in Paris had arranged for Valentine's machine to be sent to the Döberitz Military Ground near Berlin where I was to fly in front of German authorities. It was now March, and when we arrived at the ground, the wind was strong, but His Excellency General Baron von Lincker of the German Aviation Corps and his officials, were waiting for the first demonstration. I understood they were convinced our machine would not be able to fly in such bad conditions, so I thought I'd better take this up as a challenge to prove that Bristols were really good machines. I got into the Prier, staggered along a few hundred yards, rose, and suddenly shot up. After handling the machine with difficulty for a considerable time in the wind, I came in to land, but made an extremely bad one, misjudging my height and almost smashed the machine as I hit the ground and bounced badly. Without waiting for questions, I took off again knowing if I returned a little later the impression made by the shocking landing would have worn off a bit. I landed the second time, this time in much better style, but imagine my surprise when I found the German officials looking very pleased with what they had seen. Frank Coles was with us, also Mackintosh. Frank took me aside while Thurston was engaged in conversation with the officers and told me Thurston had explained that my heavy landing was a demonstration to show the strength of the undercarriage, and that this is what I always do on my first flights. That was a relief! In the following days I took up several officers and did cross-country flights with some, then they tried out the Bristol for themselves and liked it.

One day the Kaiser paid us a visit to see Prier's machine while the trials were taking place. I was not presented to him, as I'd been to the King of Spain. Indeed, quite the reverse, I received strict instructions to keep out of sight. The emperor would not be pleased to see a British pilot flying before him! Apart from this strange attitude, I saw two small incidents while in Germany through which the everyday discipline of the German was brought home to me. Now and again I went for a swim at the baths and was with Frank Coles on my first visit. We got undressed and dived neatly into the water, but the attendant rushed over in an outburst of fury. I'd never seen anything quite like it, and could not understand what it was all about, but apparently it was a terrible thing for the public to go straight into the pool without first going through the washroom which neither of us had noticed. On the second occasion, when I'd made certain I'd gone into the washroom before appearing, I saw rows of children lined up in the shallow end doing terribly strenuous exercises to order in an extraordinary regimented fashion.

The Prier was left in the hands of a German pilot, Lieutenant Macintoon. As we prepared to leave, I watched him come in from a flight. A high speed landing with the tail up was necessary with Prier's machines and if the usual three-point landing was made, the machine would start to bounce, each successive bounce getting bigger until a crash was unavoidable. The lieutenant landed too slowly. In the confusion that followed, he ran into the latest 100-horsepower Mercedes Taube monoplane, *Taube* literally meant *dove*. Most of the German planes looked like birds. The wings of the Taube were badly smashed but Coles and Mackintosh gave a helping hand in repairing the damage to both machines before we parted. Again our trip had been most successful, orders were placed and the military authorities asked for a licence to build Bristols in Germany.

We had a fine reception when we got home. The Prier machine at the Paris show was well spoken of in the journals, 'Occupying one of the central stands was the Bristol two-seater monoplane, a thoroughly worthy example of British enterprise and workmanship. None could be said to be superior to the Bristol.' We received a lot of publicity too, concerning our trips to Spain and Germany. *London Opinion*:

> 'A British pilot who has returned from aboard with an enhanced reputation is Mr Howard Pixton. He and Mr Harry Busteed have been flying a Bristol biplane and monoplane in Spain before the King and the highest military authorities, and so impressed everyone that the Spanish Government have bought a number of Bristols and have kept Busteed there as Chief Pilot at their Military School. Mr Pixton then went to Germany with a Bristol Military monoplane and did some fine flying at the German Military aerodrome at Döberitz. So impressed were the Germans that it is more likely that Germany will have more Bristol monoplanes for their Army this year than England will. Still, a prophet is ever without honour at home.'

What made our trip so important, of course, was the Prier monoplane, some thirty-four were made, single and doubles with side-by-side and tandem seating, but just at the height of success Prier decided to leave Bristol and return to France. However, we were introduced to Coanda, son of General Coanda the Roumanian War Minister, who had exhibited at the Paris Air Show and was replacing Prier as chief designer. Coanda showed promise as a sculptor but had turned to engineering, training in France where he took up aeroplane design. Though barely twenty-six, he was very highly thought of, his ideas were far advanced than most constructors and his arrival was noted with enthusiasm by the Press, 'The latest celebrity to take services with the British & Colonial Aeroplane Company is Monsieur Coanda, who is certainly one of the most original minds at present connected with aviation.'

What stunned us on our return was the news of another death during our absence. We were terribly shocked to find that Graham Gilmour had lost his life... *Gilmour dead?* It was hard to believe. He'd been the life and soul of Brooklands, a great personality, welcomed wherever he went, participating in every flying event, well-known for his love of daring and only twenty-six, the same age as Coanda and me. The accident happened in February when, after leaving Brooklands in a bumpy wind, a wing of his Martin-Handasyde monoplane collapsed over Richmond and he fell like a stone into Old Deer Park. Gilmour left a letter to be opened in the event of his death, and in it requested that there should be no mourning, but this was unavoidable, everyone was upset. He wrote:

> 'If I am smashed up by aeroplane, which is very probable, I don't wish my body to be taken (to Mickleham Churchyard, where I wish to be buried near my father and mother) by the ordinary funeral hearse, which is a beastly thing, or by train. It is, if possible, to go by motor lorry, and then if convenient by a four-wheeled farm cart. No mourning of any kind. Coloured flowers, if any. No moaning. I wish everyone to be merry and bright. If it is wished to put up a tombstone let it be a quarter scale model of the plane I was killed with in stone, bronze or iron, or model of the *Big Bat*, or other planes, it does not matter which. The remains of the plane I wish given to Clifton College. Perhaps it will be possible to re-erect it there, of course, provided they would like it. No bell tolling at the church, beastly idea. Dead March if they like, some of them are fine.'

As he wished he was buried beside his parents at the churchyard in Mickleham, nine miles south of Brooklands. Most of the Brooklands people attended the funeral, Major Lindsay Lloyd, Martin, Handasyde, Sopwith,

1914
MEN OF THE MOMENT IN THE WORLD OF FLIGHT

Howard Pixton.

The Schneider Trophy Trio and the Schneider Float Plane, now with its wheels back on, Brooklands, 1914. *From left to right*, Victor Mahl, Howard Pixton and Tommy Sopwith.

1914
THE SCHNEIDER TROPHY

The Schneider Trophy.

1910

Howard Pixton Licence No 50, 1910.

1910

Howard seated in the 35hp Roe Triplane, the 'Tripe Hound', Brooklands, 1910.

Roe crashes the 'Spare-Parts' Triplane, Blackpool Meeting, 1910.

1911

Howard seated in the 35hp Roe Biplane, the 'Pixie Plane', Brooklands, 1911.

Conway Jenkins crashes the Roe 'Pixie Plane', Brooklands, 1911.

1911

FIRST ROUND BRITAIN RACE, STAGE ONE

THE PENALTY OF FAME

Mr. Cody fondly and forcibly kisses Mr. Pixton at Hendon on his arrival. Mr. Grahame-White (in the background) appears to enjoy the joke

THE TATLER,
AUGUST 2. 1911.

1911

Howard seated in the 70hp Bristol Box Kite, the 'Bumble Bee', Round Britain Race, 1911.

Howard crashes the 'Bumble Bee' at Spofforth, Round Britain Race, 1911.

1912

Howard demonstrates the 50hp Bristol Prier Monoplane before King Alfonso XIII and officials, Madrid, Spain, 1912.

German pilot crashes the 50hp Bristol Prier Monoplane into a Taube shortly after Howard's demonstrations, Döberitz, Germany, 1912.

1912

The 80hp Coanda Monoplane being righted after Howard's wet landing patch turned out to be a pond, Bucharest, Roumania, 1912.

Howard standing, Captain Dickson seated, in the 80hp Coanda Monoplane, Torino, Italy, 1912.

1913

A Coanda Biplane catches fire mid-air, Pizey and Fellows escape death, Larkhill, 1913.

A Prier Monoplane crashes and catches fire, Major Hewetson dies, Larkhill, 1913.

1913

A Coanda Monoplane spar buckles, wing collapses and plane crashes, Geoff England dies, Larkhill, 1913.

A Cody biplane collapses mid-air, Cody and passenger fall out, both killed, Farnborough 1913.

1914

One of two touch-downs required to qualify before the Schneider Trophy Race, Monte Carlo, 1914.

The Victorious 100hp Sopwith Tabloid steeply banking in front of the Casino, Monte Carlo, 1914.

1914

A bouquet on Howard's winning plane, Monte Carlo, 1914.

'Bravo, Monsieur Pixton!' Jacques Schneider congratulates Howard on his victory, Monte Carlo, 1914.

1914

Howard in a downwind flip-over during the delivery of a 80hp Sopwith Tabloid for the War Office, Farnborough, 1914.

1917

Howard seated in a 120hp Sopwith Triplane ready for testing, AID, Farnborough, 1917.

1917

Leeds 160hp FE2b Plane, presented to Lord Desborough, Howard standing, Lord Desborough seated, Roundhay, Leeds, 1917.

1918

AID 200hp DH9 Plane, the type presented to Lord Weir, flown by Howard, Heaton Chapel, Manchester, 1918.

ROYAL FLYING CORPS

Captain Howard Pixton.

Radley and so many others that the little church could not accommodate them all.

Though the Government kept an eye on aviation, it had not hastened its development to any great extent, but things were happening now. During March 1912 Colonel Seely, Under Secretary of State for War, spoke of the formation of a Royal Flying Corps in a debate at the House of Commons. The year-old Air Battalion was to be given a new face and backing beyond expectations.

> 'There is to be one Flying Corps, embracing soldiers, sailors and civilians, all who can fly, and will undertake the obligation to serve their country in time of war in any part of the world. No man shall hold executive rank in the Flying Corps unless he is himself an expert flyer. The present Air Battalion ceases to exist, and part of it is absorbed in the new organisation. The Corps will be one Corps and, as far as possible, all the officers will be paid alike and treated alike because they will run the same risks, and have the advantage of doing the same daring deeds. In a purely land war the whole Flying Corps will be available, and in a purely naval war the whole Flying Corps will also be available. The headquarters of the Corps will be on Salisbury Plain and a large tract of land has been purchased for the purpose at a cost of about £90,000.'

A White Paper, gave full details of the new Flying Corps made up of five bodies, Military Wing, Naval Wing, The Reserve, The Central Flying School, and The Royal Aircraft Factory. Colonel Seely explained the functions of the new Central Flying School to be established at Upavon:

> 'We do not propose to use the Central Flying School for teaching officers, we propose they should learn the elements of air elsewhere, and then come to the Flying School for the more advanced course...They will learn progressive flying, mechanics and construction in all its details, meteorological observations in the air, navigation and flying by compass, cross-country flights, photography from the air, signalling by all methods and, most important, the types of warships of all nations.'

There had been anxiety concerning the Royal Aircraft Factory, a new name given to the existing Army Aircraft Factory at Farnborough which would go full ahead building experimental machines and improving established designs, until it was stressed that 'the most definite assurances have been given from the highest quarters that the factory is not to be worked in competition with the private constructing firms, and its scope and limitations have been very emphatically defined.'

And later, a motto for the Flying Corps... 'It is announced that His Majesty the King has been graciously pleased to approve of the Royal Flying Corps being permitted to adopt the motto *'Per Ardua ad Astra',* Through adversity (through struggles or toil or through bolts and bars) to the stars...

In April, I had another adventure with Burney, the Admiral's son. He wanted me to test the hydroplane which I'd seen being built at Filton in the utmost secrecy and now it was ready for its secret trials. Its fuselage was boat-shaped to ride the water, underneath were three legs fitted with a series of hydro-vanes, two in front with water propellers, and one at the rear. In theory one started the engine, let in a clutch which drove the water propellers then, when the machine lifted itself above the water on the legs, one would start the air propeller and take off in the normal way. The idea was ingenious but I had little faith in it and could see the distinct possibility of ending up in the water. We were ready to leave – Burney, Stanley White, Mrs White, a few helpers and myself – and we took the machine to Avonmouth. Here, with the help of a small crane on board, we placed it gently into the lighter *Sarah* and when done, sailing from the Bristol Channel, we arrived a few hours later in a sheltered corner of Milford Haven, the quiet, landlocked bay in Wales.

On the day of the first trials we put the machine into deep water only to discover the first problem was to get it started. The plane leaked and the water propellers wouldn't start. We towed it and they still refused to function and after several disheartening attempts, Burney had an idea. He would ring his father and ask him to send a destroyer to tow it, the propellers were certain to work if we could get up enough speed. Sure enough through influence and the position of Burney's father, a destroyer came into the bay and once again the hydroplane was lowered into the water by the crane and a thick rope attached to it for towing. When I was in the cockpit, Burney waved an axe he'd secured, and assured me that he would cut the rope if there was any sign of my turning over. My position proved extremely uncomfortable for the machine was utterly unstable on water and constantly threatened to upset me. I got soaked. For three weeks we persevered, but the experiments were a failure, we had hardly managed to lift the machine out of the water, and had we got as far as trying to land on water with its propellers buzzing around on the legs, I would have nosedived for certain. I left, glad to suffer no longer but naturally sorry that Burney's idea had been so unsuccessful. He started on an improved version but did not get anywhere with it and continued to work on his machine with folding wings for deck transport, also on one with inflatable wings which proved impracticable.

I returned to Germany soon afterwards as the result of our first visit, to the new Deutsch Bristol Werke Company at Haberstadt at the foot of the Hartz mountains where Bristol Box Kites and Prier monoplanes were being made under licence. One of my jobs was to form a flying school and to instruct six German officers. They learnt readily and the only nasty moment happened when a Cavalry officer froze on the control stick as we were coming in to land. Possibly he imagined he was on his horse with the reins taut to stop, and I had to brutally jerk the stick from his grip to prevent an accident. A few weeks later Kemp, my colleague from the Roe days, arrived to take my place. On departure the officer pupils gave me a very fine complimentary dinner, presenting me with a silver cup inscribed with their names as a token of gratitude.

Being otherwise engaged, I was not flying over to Brooklands as often as I did but on one of my visits I saw the new Avro enclosed monoplane, the first of its kind and a welcome sign since we could get very cold flying in all weathers. There'd be at least a couple of attempts made at enclosed machines during 1911, a Piggot was shown at the Olympia Show and there was a clumsy Blériot Aircar, but nothing had become of them. The new Avro flew well and being enclosed some thought the pilot's view might be restricted, but the converse applied. It was well-equipped with windows giving the pilot a much wider view than normal, but to enter the cabin one climbed in from above, and a correspondent describing this wrote, '*The pilot climbs in through a trap door in the roof, and in case of an upside down landing, makes an emergency exit through the side with a pair of wire cutters.*'

Brooklands now had competition of a sort as the London Aerodrome at Hendon, owned by Grahame-White & Co, had opened to the public at Easter and thousands turned out for its inaugural three-day meeting. Its main function was to entertain with Saturdays as race days, flying exhibitions and passenger flights taking place on Sundays, and ladies' days and fetes were on the future programme. It was described as, 'The most important and largest privately owned aviation ground in England.' Its popularity grew. So Brooklands, the Mecca of pioneer flight in Britain, was no longer the exclusive place for regular meetings but, unlike Brooklands which was primarily an experimental ground devoted to construction and testing new machines, Hendon was a show place. For me, and most of us, no place could compare with Brooklands.

Larkhill was living up to its name for spring brought out the larks in song. The air filled with starlings and flying beetles, the May bugs, which bumped into one and clung to clothing with their sticky feet, nature's clumsiest aviators. A certain pilot went frantic whenever these beetles came near him, swiping them away the best he could and usually ended by covering his face with both his hands until someone came to his rescue. Togo, Gordon England's little black Scottie, would stand under the water spouts of the sheds at every opportunity to get a shower and feed on the dead beetles washed down on rainy days. I was very fond of Togo, he enjoyed coming for rides in the Argyll as much as Whisky did at Brooklands. But one day, like me, the poor thing received a sharp nip from Gilmour's wolf-dog. I knew how it felt... Busteed had adopted Seti after Gilmour's death.

In between my overseas visits, I continued instruction at Larkhill, and with an increasing number of pupils, competition for lessons had become very keen. One of our pupils was Prince Serge Cantacuzène from Roumania heading a flying mission of officer pupils. He'd come to England especially to learn to fly and observe our methods of running a school, and was the first prince to take flying lessons at a British school. There was also a group at Larkhill who called themselves *the Terrible Five*, comprising Geoff England, at last fulfilling his ambition to fly, Robert Barnwell, brother to Barnwell in the Design Office, and three Australians, Captain Penfold, Sydney Pickles and Lindsay Campbell.

Captain Penfold, whose real name was Taylor, was a balloon parachute jumper well-known under his professional name in Australia and, although the public still thought of flying as the quickest way to the next world, I was never more surprised to find the same view shared by him. He'd had most of his bones broken and told me he'd once dropped into Sydney Harbour from a balloon and was rescued barely in time to prevent the sharks from getting him, yet, despite his dangerous life he thought flying much too dangerous and wondered that anyone had the nerve to do it. Why he wanted to learn to fly, I shall never know, but each time I took him up he must have gone through hell, even though he proved to be an excellent pupil, but when he passed his tests he hardly went near an aeroplane again. The two other Australians, Campbell and Pickles, one day literally fought to decide who would take the next lesson, and I was compelled to give them a little lecture on the prestige of being a member of the Flying School. Campbell was nearing fifty, and had been commissioned to organise an Australian Flying Corps, coming all the way from Australia to learn to fly for that purpose.

During the month of May 1912, the news that Wilbur Wright was critically ill with typhoid fever reached us, and on 30 May he died, aged forty-five. The cryptic headlines reporting his death read, '*The first aviator to die a natural death*'. His brother thought the anxiety caused by their having to fight legal battles to protect their patents had contributed to his run-down state of health and his physical fatigue which made him easy prey for the attack of typhoid that caused his death. They were in the throes of building a magnificent mansion in Dayton when Wilbur died.

Thirteen people in England had died now in flying accidents, including the first fatal accident occurring within the Royal Flying Corps which caused much concern. Captain E.B. Loraine, who was not connected with our actor friend Loraine, was killed at Larkhill in a Nieuport side-slipping on a tight-banked turn to earth. With him was Sergeant Wilson. They'd died instantly, Loraine with a fractured skull and Wilson a broken neck. A service was held at the Guards Chapel, Wellington Barracks, and a memorial cross erected at the junction of Salisbury–Devizes Road and the Amesbury–Shrewton Road, inscribed, 'To the memory of Captain Loraine and Staff Sergeant Wilson, who, whilst flying on duty met with a fatal accident near this spot on 5 July 1912, Erected by their Comrades.'

In the same month, the sad news came that Hubert Latham was also dead, not by a plane crash, but during a hunting expedition in Central Africa. He was only 29-years old. His flying career had started in 1909 not long after his Channel attempts and from then on he broke record after record, but he had not done much flying recently having decided that he would no longer take part in competitions and exhibition flights. His body was brought back to France for burial in the family vault at Le Havre.

Shortly afterwards preparations for the Military Aeroplane Trials began and Larkhill was the spot chosen for them commencing early August 1912 and continuing to the end of the month. It was to be the greatest event of the year, and much work had gone into its organisation. Jullerot, England, Busteed and I stayed on the Plain to help to make way for them, sending our pupils to the already crowded school at Brooklands and handing over some of the sheds to the War Office. Rows of round topped hangars to accommodate competitors' machines were erected, tents put up as sleeping quarters for all concerned, including the Press, and for many months constructors had been busy preparing machines to fulfil the War Office requirements. Over thirty, British and foreign, were entered. Bristol, anticipating good results with two Coanda monoplanes and two England biplanes, had once again entered the most with Busteed, Valentine, England and myself to fly them. The tests included assembly, climb, altitude, gliding angle, three-hour duration, maximum speed, minimum speed, take-off from ploughed field, pulling up on grass, rough-weather flight, and dismantling for road transport and assembly. The judges were Brigadier General David Henderson (Director of Military Training, War Office), Major Sykes (Officer Commanding, Military Wing, RFC), Captain Godfrey Paine (Commandant, Central Flying School, Upavon) and Mr Mervyn O'Gorman (Royal Aircraft Factory, Farnborough).

August 1912
First Military Aeroplane Trials

Entrants	British Planes
BUSTEED	**Bristol-Coanda monoplane**
VALENTINE	**Bristol-Coanda monoplane**
ENGLAND	**Bristol-England biplane**
PIXTON	**Bristol-England biplane**
CODY	Cody monoplane
CODY	Cody biplane
PARKE	Avro enclosed biplane
CHARTERIS	Avro biplane
SOPWITH	Coventry Ordnance biplane
SOPWITH	Coventry Ordnance biplane
FENWICK	Mersey monoplane
MCDONALD	Vickers monoplane
PETRE	Handley Page monoplane
RAYNHAM	Flanders biplane
PARR	Piggott biplane
Did not arrive in time	Harper monoplane
Not assembled in time	Aerial Wheel monoplane
VEDRINES	British Deperdussin monoplane
BELL	Martin Handasyde monoplane
PORTE	British Deperdussin monoplane

August 1912
First Military Aeroplane Trials

Entrants	Foreign Planes
VERRIER	Maurice Farman biplane
HAMEL	Blériot monoplane
PERREYON	Blériot monoplane
SIPPE	Hanriot monoplane
BIELOVUCIC	Hanriot monoplane
PREVOST	Deperdussin monoplane
MOINEAU	Breguet biplane

RHODES-MOORHOUSE	Breguet biplane
CHAMBENOIS	Borel monoplane
BLASCHKE	Lohner biplane
BIER	Kny monoplane

As my lodgings, The Crossroads, were only a quarter of a mile from Larkhill, I was very conveniently situated for the trials and thought it would be nice for Cody to stay with me and had invited him, and his family, to do so. Upon accepting, I reserved board for them and met Lela for the first time. Cody had the most unfortunate luck with the new monoplane he built especially for the trials, apparently having had second thoughts about his *Cathedrals*. The new monoplane had been built with the family's help and was immense, about 37ft long, 46ft span, 12ft high, and powered with a heavy 120-horsepower Austro-Daimler. His test flights on Laffan's Plain were encouraging, with speeds reaching near 90 miles an hour, then came disaster. During a test flight, the engine cut out and Cody was forced to glide down amongst cows grazing below, smashing the plane. Cows were always around his shed. He had no time to rebuild it and so he put his Austro-Daimler into the old *Cathedral*. Although his loss had been a heavy one in costs, he was as bright and optimistic as ever, it was a pleasure having him around.

At last the opening day for the trials came. Discounting a few latecomers, only twenty machines were at the scene on the last day of acceptance. With the absentees, it was to be a competition between British and French machines. Most of the British entries had French engines, the most popular being the Gnome, for still Britain had failed to produce an aero engine of any note. Coanda's monoplanes looked more warlike than most, with their disc wheels and aluminium sheathing covering the front part of the fuselage which gave them an armoured appearance. His work inspired greatest respect, the Coanda monoplane being one of the best built and soundest machines of this period.

First on the agenda was the assembly of machines in the quickest time. The planes were dismantled, their pieces laid out on the ground, assembly began from the moment they were picked up to the time the machines were ready for flying. Some had four men working on them, others had five or six men. The Coanda planes took well under half an hour and the England planes being big machines took longer, Cody was helped by his son Vivian and two men, they took over an hour, but the Maurice Farman with all its wires and fittings was the longest, over nine hours with five men. After assembly, a circuit flight was desired without adjustment or tests, and to see that nothing further was done to the machines each was guarded by officials. Major

Brooke-Popham flew a Government Avro over the nearby woods, setting an approximate course for the assembled machines to fly. Nothing more than the assembly tests were done during the first few days, apart from the competitors' machines being weighed. The weighbridge consisted of three platforms and the front wheels were placed on a girder resting on two of the platforms, the third took the tail, but as gusts were affecting the accuracies of the readings, the weighbridge had to be taken from the open air into a hangar. Once the machine settled, simultaneous readings were taken and, although not exactly a perfect method of weighing, a good mean was found. Readings had to be within the maker's estimates but, on the whole, they were found far in excess, some by as much as 500lbs, half as much again as the weight of an average sized machine.

During the third day of the trials, I was saddened to learn that Lindsay Campbell, one of *the Terrible Five* and the man who was appointed to form the Australian Flying Corps, was killed at Brooklands after stalling a Bristol monoplane. He'd not long passed his tests. When they found him he was sitting in his seat alive, but he died shortly afterwards and I was told that had he worn a safety helmet he would have survived as the accident was no worse than many we'd seen where the pilot just got up and walked away. Although many fine helmets had come onto the market, most of us didn't wear one. Campbell had brought his wife and two children to Britain while learning the flying business, and following his death his family was given a first class passage back to Australia, plus a sum of £333 raised by sympathisers.

The tests were temporarily suspended on Thursday, 8 August, when the trials were attended by Colonel Seely and about 150 members of Parliament and other guests, to witness, it was said, some of the world's cleverest and most daring pilots. We were asked to show our own personal skills to the best possible advantage for the benefit of the guests. Of course we did just that and our visitors left having spent a satisfactory day at Larkhill where they were able to see all the new and latest planes of the day.

The trials continued but I'd never known such an August, we were having winter weather in the heart of summer, and it was proving to be the wettest August for years, such a contrast to the hot, thundery weather of last year. In between times, Cody played cricket with the lads, using a spare Deperdussin skid as a bat, and was induced to give lasso displays. There was no escape from Cody once he had a rope in his hands since he'd run his own Wild West shows before his flying days and was an expert in such skills. Whenever possible flying was taking place between 4am and 8pm, pilots were noted as early risers to get the best of the day, so it was usual for those connected with the trials to be up before breakfast to do a test. A blue flag was flown when it was officially good enough for flying, and the results of the tests undertaken were being placed on a notice board for all to see. Geoffrey de Havilland, who'd joined the engineering and design staff at Farnborough, was flying a BE2 biplane but, although it could not be entered officially being an Army plane, permission from the Royal Aircraft Factory had already been granted for it to perform alongside us.

The stiffest part of the competitions was the three-hour duration flight which was coupled with climb and altitude tests, when we were desired to climb 1,000ft within five minutes, attain a height of 4,500ft and maintain flight at this height for an hour. A passenger, usually a military man, accompanied us and weight was added if necessary to make up a 350lb carrying load. At the same time, by filling our tanks for four and a half hours flying, the consumption of petrol and oil used was estimated on refilling. The Government supplied oil and petrol free. Passing the three-hour duration was compulsory and had to be done before qualifying for the other tests so it was all important. A few did not make it. While some pushed ahead with their tests, others waited for favourable conditions, but as early as the tenth day Prevost, flying a Deperdussin, had finished all his tests, the first one to do so, and had returned to France.

Then the competitions were marred by a death. On 13 August 1912, Fenwick of Liverpool flew towards Stonehenge and crashed and no further flying was done for the rest of the day. Fenwick's splendid but fragile pusher monoplane, *The Mersey Plane*, required, I felt, further development and should have never been allowed to enter. We were too far off to see exactly what had happened, but apparently a gust of wind proved too much for the little machine and it dived as if to get out of the air disturbance, climbed and dived again, a dive which changed into a vertical drop. I was one of the first to reach the spot, the machine was smashed to splinters, there amongst the wreckage was Fenwick, I have never seen a man so terribly injured as poor Fenwick, it was horrible, the propeller had cut off the top of his head. During our brief encounter we'd all grown very fond of Fenwick and were very upset he'd met his death in the south while flying as our guest. He was one of the leading airmen in the North, and the holder of a very early licence, No 35, and we knew him as *the Pilot in the Sand Shoes* since he spent most of his time at Blundell Sands and never wore anything else on his feet.

It was on the day of Fenwick's death that I was called away at Sir George's request to visit him at his home in Bristol. Our machines hadn't done much up to now, Valentine had walked out and Gordon England couldn't get his machine off the ground and had taken mine. On arriving at Sir George's grand home, I went to the first entrance I saw but it turned out to be the tradesmen's entrance. However, a maid took me through the kitchens and along a passage to the drawing-room where Sir George and I had an informal talk about the trials over a drink. He told me that Valentine disliked Coanda's machine and had refused to fly it. England's biplanes were withdrawn, would I take over Valentine's machine? I was only too happy to do as much as I could for the company and for Coanda's sake. Valentine had smashed the undercarriage and damaged a wing and nose of the machine, but these were attended to before I took over.

My new machine was number fifteen and before attempting any of the tests I made a few flights, but time was running out for me, the trials were half over and I would have to do some hard work to catch up. I successfully did my three-hour duration flight, taking off at 5.33am and, not wishing to fly more than the necessary time required to qualify, landed at exactly 8.33am. It was very windy and three others were also trying to get through this qualifying test during the time I was in the air, but they all gave up. By sticking it out, I was now free to continue with the other tests while they still had to face this initial test again. However, during my three-hour duration flight, though windy, the weather was so clear that I had seen the whole of the Isle of Wight 45 miles away, also shipping in Southampton waters and the Solent, thus confirming Colonel Seely's statement that at 4,000ft one could possibly survey such an expansive area.

Next, I did my speed tests and got up to a good average of 73 miles per hour, flying twice against the wind and twice with it, and easily passed since the requirement was 55 miles per hour. Without landing I flew as slowly as I dared for minimum speed which had no limit but that of safety. Reporters were everywhere.

> 'The performance of Mr Pixton, who got up a speed variation of 58-73mph, was by all accounts startling in more ways than one. The process of keeping the machine just alive by intermittent switching of the ignition is certainly an interesting and by no means uninstructive performance, whether it may be everybody's joy is another question.'

For the gliding tests we had to descend with a passenger from 1,000ft with the engine cut off, but I found the conditions unsuitable from the expected height and got no marks for this exercise. Hardly a thing went unreported.

> 'Early in the afternoon, Mr Pixton with Lieutenant Ashton as passenger, started out for his gliding test, but came back saying he found it bumpy. This is from our Champion Wind Fighter!'

Ease of steering was demonstrated on the ground by circles and figures of eight, and range of view was done by officials on bad flying days in a hangar with its floor marked like a chess board. An estimated range was calculated by the number of squares seen from the pilot's seat with allowances made for the varying heights of the aircraft. The Coanda topped the list. Then there came my ploughed field tests. Sippe had succeeded in bringing his Hanriot down on its nose in a field while Busteed landed in a haystack, of all places, in an adjoining field during this test. I landed all right but failed utterly in taking off due to no fault of mine. It had poured with rain just after I'd applied for the test to be observed and the field was reduced to a quagmire, the wheels spun uselessly. I just couldn't get a grip, but I don't think any machine could have risen from that mud. Cody had had no problems at all with his *Cathedral*, having taken the test during a fine spell. He teased me about it, couldn't I pick a better day?

I took the normal landing tests with equal failure, for what was required was a stop within 75yds and I far exceeded that. Once again Cody had done a fine job in stopping within 33yds, he had his own ideas about braking, a chain which looped around the front skid acted as a brake when he pulled a cord. By miscalculating the length the machine would run, I found I had landed too near the hangars and was going straight towards them. As I realised I would not be able to stop in time I took a chance, opened up the engine, aimed for the Sun Gap, got through but came face to face with a fence at the back. This was bad flying, the strength of the machine surpassed my judgement and with only a second or two to act, I held the nose down until I'd almost reached the fence, pulled up sharply, just scrapped over and touched down on the other side, like a horse taking a hurdle, and there alarmed some spectators who scattered in all directions on finding me suddenly charging at them. No one was hurt, I'm glad to say, and a write up on this episode was headed, '*Aerial Fence Leaping*'.

What I excelled in though, was the rough weather flying test, and for that it was necessary to make a short flight in a wind averaging 25 miles per hour. When I took my machine out however, with Captain Hamilton as passenger, the wind was rising and topped a force of 47 miles per hour during the flight. It aroused a great deal of comment in the aviation journals and flying circles both here and in France since it was a new world record in wind flying, and charts of my flight readings and Latham's 25mph record in 1909, were reproduced.

'Today it is Pixton on the Bristol monoplane that has set tongues wagging. Pixton and Captain Hamilton suddenly surprised the very limited field by going up in number 15. For a quarter of an hour Pixton fought the gusts, which ranged from 17mph to 47mph in velocity. A really remarkable flight. No doubt it was an exhibition affair designed to take the shine out of the Frenchmen.'

Shortly afterwards I received a surprise telegram from Vienna. *'Wien. 16 August 1912. Pixton Pilote. Mes meilleurs felicitations, General Coanda.'*

The trials neared an end but there was one incident to come which was to go down in history. It concerned Parke who, having been held up during the trials, was doing his three-hour duration flight on the enclosed Avro the day before the competitions ended. He and his passenger were just finishing their flight and coming in to land when it was feared there was going to be a second fatality. In beginning a spiral glide to earth from 600ft, he suddenly went into a spinning motion with the machine 'hanging on its tail' and rapidly falling in a manner which was comparatively unknown in aviation, but he made a miraculous recovery from what appeared to be certain death when only 50ft above the ground. Though naturally very shaken after his alarming experience, he explained to us what had happened, but was just as puzzled as we were.

> 'I'd glided down some distance when I thought the machine was much too steep and not banked sufficiently for the turn I was making. I therefore pulled the elevator back and touched the ailerons, hoping to reduce the depth of glide but found I was still pointing downwards, and the machine was turning round and round at the same time. It was frightening, and I had no idea what was going wrong. I hardly remember what I did, it was out of control. I opened up and pushed the rudder to turn inwards to promote a dive but kept on spiralling, then I booted the rudder over in the opposite direction and found I'd straightened out.'

I was one of many who did not actually see Parke spinning, but I recalled how I'd seen Raynham turn somewhat similarly but less dramatically. I'd not given it much thought until now. Parke was certainly lucky to be alive and his experience must have been all the more frightening in an enclosed machine, a type which none of us had tried. The dive caused a great deal of excitement. 'Let no one forget the rule to rudder outwards from a spiral dive that has already acquired a high velocity.' There was a lot of talk about this dive, but no one had the answer. From then on the spin was known as *'Parke's Dive'*.

The trials came to a close on Tuesday 26 August, with Parke and Bell undergoing their speed and transport tests. Road transport, the last of the tests, was when machines were dismantled and packed for the road, towed away and observed as they went uphill and downhill, returned to the sheds, assembled and flown.

Most of us had suffered some misfortune or another during the Trials. Avro had not progressed well because of delays but was the only all-British machine to complete the tests, the Handley Page didn't do much, nor did Vickers, nor Martin Handasyde, nor the Coventry Ordnance, a company which had taken over the Howard Wright concern. Their machines were to have been flown by Sopwith, but he had to leave for America during the first week to take part in an International Boat Race. Raynham, who was also flying the Flanders Biplane, had taken his place but failed to make much progress on any machine. I thought Sippe's Hanriot had done exceptionally well, as did most people, and was to my mind one of the best. Blériots had done quite well, but they were no longer the fastest machines available and were well under that of Bristol's, Deperdussin's and Hanriot's speeds, a change from a year or so ago when they took most of the records and prizes.

The results! And who won the main £4,000 prize? Why, none other than Papa Cody on his old faithful, with Provost second. Not only did Cody take the International Prize of £4,000 but £1,000 also, as the National Winner. He was jubilant! His *Cathedral* had surpassed the best machines established firms could muster. What history! The Government discovered the best machine for performance was not an up-to-date one but one whose design had to be rejected for reproduction! The whole point of the trials had been to find machines good enough for the Government to place orders with their companies. Nationally Busteed, Bell who had changed planes, as I had, and myself came a joint third, there was no second prize... So ended The Military Trials, with Coanda machines coming off very well, and Cody's as the best but unwanted.

Then came the reasons. The Judges' Committee placed the most desirable requirements in order of importance as being, first and foremost, Speed and Flexibility of speed, then Climb, Glide Angle, Landing within a short distance and Good View. High value was not placed on an aeroplane to withstand rough weather, so my record did not amount to much, but I thought any importance attached to assembly and road transport should be discounted now we were more advanced in design, as machines were made to be flown and not packed. The committee was of the opinion that Fenwick had endeavoured to meet the requirements of war with his *Mersey Plane*, the very machine I'd considered unsuitable for entry, the one considered advantage they said, being that Fenwick's monoplane had been a pusher. It was known that the Army favoured pushers in case of war since guns could be fired from the front without the propeller getting in the way. Surprisingly the final verdict of the Judges Committee had been, '*No really Military Aeroplane entered*'. So ended the greatest aviation event of 1912!

August 1912
First Military Aeroplane Trials British Winners

Prize	No	Pilot	Plane
£1,000	31	CODY	Cody biplane
£500	15	PIXTON	Bristol monoplane
£500	14	BUSTEED	Bristol monoplane
£500	21	BELL	British Deperdussin monoplane

August 1912
First Military Aeroplane Trials International Winners

Prize	No	Pilot	Plane
£4,000	31	CODY	Cody biplane
£2,000	26	PREVOST	Deperdussin monoplane

Not long after his success, Cody received a message of congratulation from King George. Although the Press had frequently called him Colonel Cody, confusing him with Buffalo Bill, he'd never accepted the title until the king called him Colonel Cody. 'His Majesty referred to me as "Colonel Cody", so in future "Colonel Cody" it has to be.'

In acknowledgement of his success, Cody's effigy was added to Madame Tussauds waxworks collection, a memorial stone, too, in remembrance of Fenwick's tragic end at the trials, was placed at South Shields High School, 'Robert Cooke Fenwick... A pioneer of aviation in England and an old boy of this school who lost his life on Salisbury Plain on 13 August 1912, while flying his *Mersey Plane*, a machine designed and built by himself, and the first to fulfil the requirements of the British Army in time of War.'

Coanda was very pleased with the performance of his planes and presented me with an enclosed cigar holder, into which a cigar could be placed for all kinds of weather and a silver cigarette case inscribed, '*To my friend Pixton. In remembrance of his Splendid Flying on my first monoplane, Military Trials, August 1912. Henri Coanda*'.

Following almost immediately after the military trials were two weeks of Army manoeuvres in Norfolk and Cambridgeshire involving simulated armies, and the aeroplane was for the first time seriously taking part, participating beside the Government airships, *Beta*, *Gamma* and *Delta*. The manoeuvres, however, began with tragedy. *Four Flying Corps men lost their lives!* The first accident occurred in the first week of September and the second four days later. Captain Patrick Hamilton, who'd flown as my passenger in the trials, and Lieutenant Wyness-Stuart were using the Deperdussin monoplane which had won second place internationally, when the machine collapsed over the Willian and Graveley area. It was thought that the engine had shifted. Captain Hamilton had only received his certificate this year, and had flown Deperdussins for a long time before his tests. He'd liked them so much that he never wanted to fly any other machine.

The second accident was particularly distressing to me and involved Lieutenant Edward Hotchkiss and Lieutenant Claude Bettington, both of whom I knew well. Hotchkiss had not long taken over the Bristol School at Brooklands as manager and chief instructor, but being attached to the Special Reserve of the Royal Flying Corps, he was flying during the manoeuvres. I was at Larkhill when he and Bettington took off on the Coanda monoplane that Busteed had used at the trials which had been bought by the War Office, and standing near me was a mechanic. To my great surprise he shook his fist at Hotchkiss's airborne plane, saying, '*I hope you break your bloody neck*'. I told him I had always found Hotchkiss remarkably pleasant, that everyone liked him. '*Well, I don't,*' he said. Not long afterwards I heard that Hotchkiss and Bettington had crashed at Wolvercot as they headed for Cambridge. A quick release clip for bracing wires to a wing unfastened, consequently cables hung loose and pierced part of the wing fabric, stress fell onto the other wing and the plane dropped to earth. On hearing the news I searched out the mechanic who'd shown such ill will towards Hotchkiss, and told him '*They're both dead!*' and he bluntly replied, '*I'm glad.*'

Although Hotchkiss's death was considered accidental, I had the nagging thought that it might have been caused by intentional carelessness by the mechanic. He'd checked the machine before they took off, but there was no evidence that this man was a menace to the safety of Bristol pilots so I decided my suppositions were best forgotten. The Public Safety & Accidents Investigation Committee recommended that, 'If quick-release attachments are used on aircraft, they should be so contrived that they cannot release themselves by vibration'. Because of these deaths within the RFC, and so soon after the trials, an outcry started. *'Are monoplanes safe?'*

The newspapers and journals began to get bits of information concerning a monoplane ban. 'According to the aeronautical correspondent of the *Daily Telegraph*, the War Office authorities have definitely decided that so far as the work of the Royal Flying Corps is concerned, there shall be a permanent ban on the monoplane!' *An astonishing statement, but to our surprise the banning of monoplanes was true*! The Royal Flying Corps was serious but it turned out that the ban was only provisional until a report was received from an appointed commission. It was not a popular investigation as it gave the impression that biplanes are safer, this surely was not so. What carried a lot of weight was twelve lives had been lost on monoplanes this year and none on biplanes, but during the year more monoplanes had been built than biplanes then, before September was out, there was yet another monoplane death, Henry Astley crashed his Blériot in Belfast. He'd got his licence about the same time I did at Brooklands and had been in many lucky smashes, this time he was badly injured and died during an operation to the brain.

Bristol was no exception in favouring the monoplane, having produced the very successful Prier and Coanda, and even while the monoplane ban was on the Coanda was much talked about, being rated as one of the leading machines of the day, but because of the controversy and the fact that Bristol was selling to the British Government, Coanda's thoughts now turned to building biplanes.

Coanda had worked exceptionally hard during the short time he'd been with Bristol and, mainly through his father's influence and his position of War Minister, we were to demonstrate the Bristol Coanda monoplane in Roumania. I was asked to go, so it was back overseas for me. During October Thurston and I prepared to leave, accompanied by Coanda himself. The machine we were to show off was a two-seater tandem, the rear seat being the pilot's, the front the passenger. A feature I liked very much about the machine was its wheel-shaped control column which, working like a stick column in principle, was pulled back to ascend and pushed down to descend and when turned it operated the warp. It was like driving a car where to steer is to maintain control. I'd first encountered it on Roe's biplane.

After a very long but pleasant journey we arrived in Roumania. General Coanda received us and we were introduced to Madame Coanda and Coanda's two sisters who, unlike their father, spoke English. They had a beautiful home in Bucharest and what charming people they were. A few days later General Coanda took us to the Cotroceni Aerodrome where the Roumanian Government had its flying school. Army manoeuvres were on at the time, and part of it was used by the Aerial League founded by Prince

Bibesco, himself an owner of a Blériot. We were introduced to the prince and chiefs of the military forces, General Boteanu, Colonel Gradisteau, Colonel Haralambie, Colonel Radeau, Colonel Teodoresco, Major Macri, Director of the Flying School, Major Filitti and many more. The Roumanian Army had a few Henry Farmans, two Blériots, a Nieuport and a Morane at the school, and I flew with several of their officers, Negrescu, Protopopescu, Parvulescu and others, during the course of our visit.

I was flying with one of the officers while the manoeuvres were on and was coming in to land, preparing to do one of my usual glides from 1,500ft with the engine stopped. I looked down and noticed what I thought was a sheet of shallow water, a wet patch after a shower, and did not take much notice of it. So I comfortably glided in, made a perfect landing, finishing with a straight run onto the wet patch. Imagine my surprise when a large quantity of water and mud splashed around me and my added surprise when the machine shot onto its nose and turned completely upside down. The wet patch was a pond! Here I was sitting upside down with my head under water pinned down unable to move with the back of the seat on my neck, in the mud unable to breathe and thinking, *What a silly death, suffocated by mud in 2ft of water!* Somehow I freed myself and after what appeared far longer than any man could possibly go without air, I got through the mud to the outside of the machine and was standing up. I leaned on the machine, recovering, Coanda would have seen it all, what a show! I couldn't have let down Coanda and his father more and deserve to be sacked! Won't be surprised if I am!

Suddenly I remembered, *Hells Bells*, my passenger! I scrambled through the water intent on lifting the front of the machine to give him clearance and air, but without success, it was too heavy and waterlogged to budge, he must be dead by now. If he were, I would never fly again with a passenger, *never!* I stumbled to the passenger cockpit bent on tearing open the fuselage with my hands and getting at the poor fellow... was there a remote possibility he might be still alive, hardly, with his head pressed into the mud, then I heard a tapping noise, what a relief! Suddenly people were all around us, up to their knees in water and in next-to-no-time we'd lifted the machine up and stood it on its nose. The trapped officer was freed, and none the worse for his unfortunate incident. They explained to me what happened. 'He hung onto his seat as you were turning over and, being smaller than you, managed to twist himself around so that he ended up crouching in the water with his head touching the floor and was kept alive by the trapped air.' Had he not twisted himself around, I would undoubtedly have been responsible for his death.

Thankfully, he bore no ill feelings towards me, in fact, although we had a language barrier, we stood in the water laughing, both realising how lucky we were, but his handsome pale-blue uniform, oh dear, it was ruined. The officer was a qualified pilot, but apparently he'd had a few smashes himself because when General Coanda saw us both dripping with water, he smiled and pointed to him, saying *'C'est Pixton deux.'* My very bad piece of piloting went unmentioned, and after the machine had been cleaned and dried out, I was flying again.

My eye caught a very weird tail-first machine made and owned privately by Aurel Vlaicu. Vlaicu was not a military man, but he used the aerodrome. We became friendly and I watched him wheel out his invention, it was one of the quaintest machines I'd ever seen, the Roumanian Cody. *'Does it fly?'* 'But of course, Monsieur Pixton.' Vlaicu made a flight for my benefit and it flew well. Canvas was stretched across a framework of cycle tubing, and the wings were completely flat, and were the only ones I'd ever seen that did not have a trace of curve on them.

Though Roumania was a sophisticated country and had a language of its own, the upper class spoke French and many newspapers were in French, but the rural people were quite divorced from the rest of the community and, both in ways and customs, they were peasants. The countryside was very flat, its fields divided by dykes, and grazing near the aerodrome were a number of buffaloes of a small species. I met some of the Roumanian country folk one day when I was forced to land some distance away from the aerodrome. It was during a duration flight when the engine cut out, and in looking for a suitable landing field I decided to make the best of a little field which threatened to cause me trouble, but there was no alternative. I brought the machine over the hedge, its wheels brushing the growth, touched down and stopped just a few feet from a ditch on the far side. It would have been hopeless trying to fly the aeroplane out again even if I were to attend to the engine, so I decided to leave the repair work to the mechanics and get them to tow it back and seek my way on foot to the aerodrome where the Roumanian officers were awaiting my return. It seemed that I was a long way from anywhere, I could see no houses, no people nor animals, but I detected a humming kind of noise, and could not make out what it was. Suddenly, from every side, peasants appeared in white, flowing robes like sheets. They must have come to investigate. As they advanced upon me barefooted, I was filled with a great deal of curiosity and perhaps just a little nervousness. *'What will they do?'* The noise, a perfect babble as far as I was concerned, was deafening as they swept around me like a swarm of locusts and climbed on the machine, swung on the wings, clambered all over the fuselage and into the seats, *'Get off, Get off'*, my efforts were useless. They only eyed me with curious inquisitiveness and took no notice. I gathered they'd never seen an aeroplane before, at least not one so close, and took advantage of it being in their vicinity, but what was I to do? They could tear it to pieces.

Miraculously, a Police official arrived on the scene, and no policeman was ever more welcomed for the crowd went silent and stood back, Thank goodness! I was getting worried. He could not understand me, but I managed to make him see, with signs, that I wanted the plane towed back to the aerodrome because he sent off a young lad at a run, and after what seemed an age, the lad returned with help. Watched by the silenced onlookers, the machine was soon ready for towing, and as it moved off it was a sight I would never forget. The machine acted as a magnet to the crowd who followed it at close quarters all the way to the aerodrome.

We stayed over a month in Roumania and fortunately I did a great deal of flying which showed the true nature of Coanda's machine, and the Roumanian Government was sufficiently interested to buy a number of them, placing an opening order for ten. Coanda, of course, was very pleased with the news. Nearly every evening of our stay, General Coanda invited Thurston and me to his home. I'd not known such wonderful hospitality, and during a visit Coanda's sister, Marie, gave me a copper plaque of the Virgin Mary to fix to my plane, as a protection against weather conditions. Roumanians made splendid flyers since they were not excitable, but their affection exceeded no bounds. When Thurston and I were leaving, several of the officers came to see us off at the station, and I bolted into the train, hoping I didn't appear rude for they all wanted to kiss me farewell. Coanda did not return with us as he was going to spend a little longer in Roumania with his father and family. Back home, I spoke to C.G. Grey about the Roumanian trip, only to find he'd included this part of my conversation in the following week's edition of *The Aeroplane*. 'Mr Pixton says the Roumanian officers have the makings of very fine pilots, being full of pluck and initiative, to which qualities is due the fact that practically all their machines have been smashed up.'

Shortly after my visit to Roumania, I married Maude Hallam, daughter of the late Chief Constable of Salford. We were married on 19 November 1912 at St Annes-on-Sea, and spent part of the honeymoon at Brooklands, Hendon and Larkhill. The second half was to be spent in Italy when during part of the time I would be flying for Bristol. The aeroplane magazines did not let us off.

> 'The news came very much as a surprise to most people, for Mr Pixton never told even his closest friends that he intended to get married till two or three days before the wedding... Another of our well-known

Aviators has renounced his bachelor freedom... I am sure our readers wish them all the happiness the world can give...'

Just before Christmas we were very shocked by the news of Parke's death, only four months after he'd miraculously escaped with his life by coming out of the fatal spin. Not long after leaving Hendon he was seen turning back with a failing engine, but he did not have enough speed in which to exercise the turn and nose-dived to earth. With him was Arkell Hardwick, manager of the Handley Page Company. Both men had to be sawn out of the plane, a Handley Page monoplane, but Parke was dead and Hardwick, still breathing, died shortly afterwards. Parke was the son of the Reverend A.W. Parke, Rector of Uplyme, and a stained glass window was erected in Uplyme Parish Church in his memory.

On Christmas Eve Edward Petre, one of the Petre brothers known as Petre the Painter, was also killed. He died on a Martin & Handasyde Antoinette-type monoplane when both its wings broke off. On the day of his accident he was making a non-stop flight from London to Edinburgh, all set to do the journey by Christmas Day, and had left Brooklands when it was rather windy. With 100 miles to go he was forced down because of an increasing wind, and met his death at Marske-by-the-Sea while trying to land. Edward was the second son of Mr and Mrs Sebastian Petre of Tor Bryan, Ingatestone, Essex. The great Petre family, of which the well-known Catholic peer Lord Petre of Thorndon Hall was the head, had lived at Ingatestone for many generations. The funeral took place at Fryering near Ingatestone while his brother, Petre the Monk, was on his way to Australia to form the Australian Flying Corps.

It was not the first time an Antoinette had collapsed. In France in 1910 Laffont and Pola were the earliest victims when their machine flew to bits in the air and a photographer had taken a picture of the actual moment it broke up. The Antoinette and its versions were really good engineering jobs apart from structural weakness of the king-post onto which the wire bracing was attached. This was inclined to bend, then the wings folded around the machine when the strain went beyond a critical point. All monoplanes were braced, but an early version of the Antoinette designed in 1911 was the very first machine to have cantilever wing spars, so doing away with the wire bracing. It was a remarkable machine with enclosed wheels, a machine years before its time, too revolutionary to be successful. Shortly afterwards the firm closed down.

Martin could not accept structural weakness as the cause of Petre's death, just as Blackburn couldn't at the time of Oxley's death and would not credit that eye witnesses had seen the machine break up. Martin and Blackburn both made good machines but every constructor had to face the reality that structural weaknesses were liable to show up in their designs with new powerful engines and increased speeds. Mr Martin complained bitterly about a report from the Public Safety & Accidents Investigation Committee issued by the Royal Aero Club:

> 'From a Committee consisting of rival manufacturers, who naturally know their business we would have nothing to fear, but from one composed, with two exceptions perhaps, of gentlemen who are not engineers, and who certainly know nothing of practical construction, we are pained, but not very surprised, to have to put up with a report as nonsensical as it is unjust.'

The year 1912 had taken a large toll of pilots, many of whom were my friends, and each death started afresh the letters of complaints from the public, *The Killer Aeroplane... Ban the aeroplane altogether...* We knew that we could be the next to go, and several times during the year I thought, as I'm sure many others did, *Well, I'm still here, but for how long?* Death was a constant companion. Each time we stepped into a machine we were aware of the risk involved, a risk which we were all prepared to take. In Britain, or connected with Britain, there were two deaths 1910, seven deaths 1911, and seventeen deaths in 1912, totalling twenty-six. *Who's next?*

At the end of 1912 I was in Italy for Bristol before filling in as School Manager at Larkhill. The Italian Government had placed a large order for about sixty machines, having increased the figure of an original order for thirty-seven. Though advanced in aviation with nearly 200 trained pilots, they felt they did not have enough machines at their disposal, and we were to demonstrate the Coanda monoplane on their decision that it was the best machine available for military purposes. Several of us went including, Mr White Smith, Secretary of the Company, Captain Dickson, Farnell Thurston and, as planned, my wife accompanied me. We did quite a bit of sightseeing en route, and on nearing Italy we took a trip up the Splügen on the funicular railway, then went through the pass to Milan where we paid a visit to the Milan Cathedral, thus completing a very pleasant journey before getting down to business in Turin. Once in Turin we met up with our agent, Mr Rose, and went to the Mirafiori Aerodrome not far from the city where the Coanda monoplane was waiting for me.

Snow lay on the ground as I set off over the Alps on a test run. In all my life, I'd never seen such impressive scenery, it was beautiful. I demonstrated the machine before Colonel Moris, Commander General of the Italian

Military Air Corps and Major Douhet, Commander of the Aviation Battalion, both of whom were highly impressed when, with Captain Dickson as my passenger, I flew at 72 miles per hour and climbed 3,300ft in 13 minutes. We were told that we'd broken Italian records for both speed and rate of climb. Dickson was still not a well man, and this was his first flight in Italy since his collision at the 1910 Milan Meeting. However, we did further demonstrations together, then I took several officers up and a few pilots tried out the machine. The consensus of opinion was, '*Without exception, we've not mounted a steadier or more breath-taking machine than the Coanda*'. One evening we were invited to an opera as guests of the Caproni Company, the only large manufacturers in Italy of aircraft and with whom arrangements had been made to build Bristols under licence. Their works area was on the outskirts of Milan. At the dinner which followed we were offered octopus, but I didn't venture to taste it. Pizey and Sippe arrived to take over. Sippe, who'd been with Roe, joined Bristol shortly after the Military Trials and was a very able pilot. They were to carry out delivery tests on about twenty machines which were on their way from Filton as the direct result of our visit, and were to superintend the construction of fourteen Bristols at the Milan works.

So, with our part done, we left Italy after a most enjoyable stay and went to Paris. There, Dickson took us to the Opera Comique. He enjoyed opera immensely, but particularly so on this occasion since he knew the leading lady.

In January 1913, my wife and I, with Thurston, proceeded to Spain, leaving Mr White Smith and Captain Dickson in Paris. I was to fly before the king again. During our first visit, Busteed and I had demonstrated the Bristol Box Kite and the Prier, but this time we were showing off the Coanda, our latest design. On arriving in Madrid we went straight to the Hotel Ritz, and after settling in took a walk, but as we were going through a park we were amazed to see a duel about to start between two men, each holding pistols. I thought duels had long ceased being a way of settling arguments, but as we stood watching and wondering whatever would happen, the men decided to apologise, shook hands and walked away with their seconds and the official in charge. The news that we were on a Flying Mission to Spain somehow got around the hotel, and an American couple, apparently interested in flying, approached us. In the course of the conversation which ensued the wife said, much to my amusement but much to the bewilderment of her poor husband, '*I do so admire you birdmen. It must be awfully dangerous. I do wish my husband would take it up*'.

After a day or two, Thurston and I made our way to the Cuatro Vientros Aerodrome and were warmly received by King Alfonso. This time we met several members of the Royal Family, including Prince Maurice and Prince Leopold of Battenberg, the queen's brothers. The queen, known as Ena, was the granddaughter of Queen Victoria and daughter of Princess Beatrice. I had the pleasure of meeting them and giving Prince Leopold, a very charming man, his first trip in an aeroplane. He was slightly lame. He flew with me for twenty minutes and liked it, and Alfonso himself an experienced pilot, also flew with me for about fifteen minutes. He spoke perfect English. At home, Prince Leopold's flight was received with great enthusiasm.

> 'What is believed to have been the first flight by a British Prince was made on the 22 January 1913 at Madrid when Prince Leopold of Battenberg accompanied Mr Howard Pixton during a twenty-minute flight on a 80-horsepower Bristol monoplane. The honour of taking the first Royal passenger is one of which the Bristol Company may well be proud.'

In all, the demonstrations went off very successfully. Orders were placed and a Spanish Commission was to visit us at Larkhill in the spring. Unlike the Italians, they did not have many pilots trained, about fifteen or twenty at the most...

Bristol was very pleased by the results of the overseas visits.

CHAPTER NINE

With Bristol as School Manager

Coanda's first biplanes were already on the scene. One was put on display at the 1913 Olympia Show where a Vickers pusher biplane fitted with an automatic gun was also exhibited, the first serious sign of a fighting machine. The ban on monoplanes had been lifted as the committee looking into the Royal Flying Corps accidents had come to the conclusion that monoplanes were not necessarily less safe than biplanes, but the effect of the investigation would leave its mark for a long time to come as many firms, like ourselves, were committed to the construction of biplanes. I took up the position as School Manager.

Bristol, now one of the largest manufacturers of aircraft in the world, if not the largest with a staff of over 400, had raised their capital to £250,000 to meet the increased demands for machines at home and abroad. They were doing a fine job and in February laid out the facts and figures of the company in a full page advertisement from which the following extracts are taken:

'The British & Colonial Company Ltd, is a private limited company started originally with a capital of £25,000, which was subsequently increased to £100,000. This amount having been fully paid up and expended, a further increase to £250,000 has recently been found necessary to cope with its ever-increasing business. The only shareholders in the company are the directors, Sir George White, Bart, Mr Samuel White and Mr G. Stanley White. Within three years, this company, established in May 1910 by Sir George White, has by soundness of design and construction so won its way to the forefront of aviation in this country... In addition to the British War office and Admiralty, the Governments of Russia, Germany, Italy, Spain, Turkey, Roumania, Bulgaria and Australia have placed large orders, in every case stipulating for tests of the severest character. So successfully have these tests been complied with that repeat orders are constantly received. The company at this moment is building for the War Office the machines designed by the Government technical staff at Farnborough, and also have in hand repeat orders from a number of foreign Governments, including Italy, Turkey, Roumania and Spain. The present output of Bristol aeroplanes is at the rate of 200 per annum, and when the extensions of the works now being carried out are completed next month, over 300 aeroplanes will be constructed a year.'

A fine company, but it seemed strange that just more than a year ago Britain was placing orders with France, whereas now we had foreign governments ordering from us. Bristol's workmanship in aeroplane construction was of such high quality that sayings *'Finished like a Bristol... Ship shape and Bristol fashion'* could well apply to these planes. Prince Cantacuzène of Roumania, who secured British Licence No 367 at the end of 1912, was staying on as a guest of the company to take delivery of some of Coanda's monoplanes, part of an order placed by the Roumanians. He'd been joined by Major Macri, Negrescu, Christescu and Parvulescu, who'd left for England to assist in the construction of these planes at Filton. Lieutenant Negrescu had also secured a British licence, No 383, during January. A second commission was due, formed by Andre Popovici, Beroniadi and Pascanu. They too, were to secure British licences 528, 542 and 543, with Bristol.

In March 1913 the first four of the Roumanian machines arrived at Larkhill. I took one up for a few minutes, but found the weather rather rough and windy. The prince, however, was anxious to get the tests completed as quickly as possible and urged me to continue. I declined. The weather was unsettled and I would not get the best results from them under such conditions. It was impractical to make further tests that day, a waste of time, and I would only take them out again on another day until I was satisfied they had undergone their tests thoroughly. He wanted the tests completed within the next four days. There was no need for haste and I told him I would certainly not let any machine go through my hands inadequately tested. Standing beside me was young Geoff England and, having been a qualified pilot for six months, was enthusiastic to make good and wanted as much flying experience he could get, and asked to do an hour's duration on one of them. I saw no reason why he should not. I'd flown as often as possible during my Brooklands days and was glad to see Geoff so keen, knowing all flying helped one to become more proficient. Though only twenty, he was already a very experienced pilot, so I did not stop him from flying on this particular day and told him to keep an eye on the weather.

I was not worried in the least when Geoff took off but, after forty minutes flying at 600 feet, I happened to look up as he was coming in to land and was horrified to see the back spar of a wing buckle, and watched helplessly as the air pressure forced the wing down so it wrapped itself round the undercarriage. The machine fell from about 60ft to the ground, and grief stricken, I rushed to the spot but there was no way of helping Geoff. Every bone in his body was broken. He was still sitting in his seat, his legs crushed underneath it, his skull fractured. His brother Gordon England had left the company and was not present when he was killed. He'd gone into partnership with Radley after having a heated row with the company over the testing of prototypes, believing they should undergo more vigorous tests before they went onto the production line. Because of his concern, he'd arranged for Geoff to work for Martin & Handasyde, but Geoff decided to stay with Bristol. It was most upsetting. Jullerot and I attended the inquest at Bulford Camp, and the question as to whether the company had ordered that the machines should be tested within a few days was raised, but no such request had been made and the verdict of Accidental Death was recorded. Geoff was now the twenty-ninth person having British licences or connections with Britain to die while flying.

One day, at the end of May, I was sitting outside the clubhouse at Larkhill idly watching Pizey and Fellows, the club's secretary, on a Coanda biplane. They were cruising around at roughly 1,000ft when suddenly, thick, black smoke billowed from their plane.. *they were on fire!* Fellows was forced out of his seat as flames shot from around the carburettor, and miraculously clung onto the fuselage. He clung on for dear life as Pizey went into a fast, steep dive before the worst happened. As soon as the wheels touched down, both men flung themselves off the machine which ran on until it came to a halt, then the petrol tank exploded and the machine burnt itself out. Thankfully, neither was seriously hurt, but watching with me was Major Hewetson, a Royal Field Artillery officer of the Indian Army who'd just arrived for a course of flying while on leave. It was the first time I'd seen a machine on fire and hoped I would never see anything like that again, and said so. I can't imagine anything more horrifying than being burnt alive. A few weeks later, in July, I did see another plane on fire, and worse still it was Major Hewetson's. I'd flown a lot with him during his tuition and on this day he went up for his solos in a Prier monoplane to qualify. I kept my eye on him as he flew, but saw him turn too sharply, stall and side-slip. I hoped beyond hope that he could recover, but within seconds it was all over. He came down with the engine full on, dived into the ground and was killed instantaneously as the tank burst and the machine caught fire. At the inquest at Bulford Camp it was suggested that a man of Hewetson's age, approaching forty-five, was too old to learn to fly. I was inclined to disagree. There were many flying of Hewetson's age and one of my finest pupils, General Henderson, was fifty when he learnt, about the same age as Cody who'd flown constantly for many years.

With Hewetson's death about thirty-five men with British connections all of varying ages, had now been killed by the aeroplane, Brooklands (5), Napier, Fisher and passenger, Campbell, and a passenger of Gordon Bell's, Hendon (2), Benson and Cammell, Larkhill (5), Captain Loraine, Wilson, Fenwick, Geoff England and Hewetson. Others had died at meetings, abroad, in other parts of England, or were lost at sea. But worse was to come!

Not long after Geoff's tragic death and three weeks after Hewetson's, another very personal blow... Cody! *Cody met his death!* So many people had been killed recently in flying accidents, but it was the last thing I imagined could happen to Cody. I was terribly upset when I heard the news and it took me a long time to get over the shock. It happened a few days before the start of the Round Britain Seaplane Race which he was entering. It wasn't due to bad piloting, but to structural weakness. He was testing his new 100-horsepower competition machine without its floats at Laffan's Plain, and flying with him was his passenger for the contest, the well-known cricketer W.H.B. Evans when, without warning, the wings of his machine folded up. Cody and Evans dropped out of the machine as it turned upside down and they fell helplessly through the sky to the ground without a hope of surviving. The machine crashed into the trees. Although I didn't see it, I vividly imagined what had happened, *two bodies falling from the sky...!* What a dreadful death! What a dreadful mental picture! Cody was only fifty-one. The Accidents Committee was of the opinion that the failure of the aircraft was due to inherent weakness. Unfortunately only just over three months previously a young officer, Lieutenant Rogers-Harrison, was killed while flying Cody's *Cathedral* which had been bought by the War Office as a result of the 1912 Military Trials, the first fatality on a Cody plane which had also broken up midair.

Of all the names connected with flying, Cody was the most famous. *Cody is dead!* Tributes were paid to him by all the newspapers, for he was a wonderful man with a wonderful personality, a self-made man without an enemy in the world. A magnificent flyer. Papa Cody! *Britain's Father of Aviation*, they wrote. He'd regarded me with feelings of friendship from the first time we met, and for no one in British aviation have I felt so much affection. C.G. Grey of *The Aeroplane*, one of his greatest friends also, wrote, '*Those of us who knew him best, loved him best*'. This was endorsed by the following description:

'Cody was a Big man. Everything about Cody was big. Cody could no more have built a small racing monoplane than he could have let petty worries trouble his big nature. A big man, he built a big machine. He wore big boots, a big hat, wore a big smile, and had a big heart, a heart so big that it must have found trouble to accommodate itself even in his big body. Nobody but those in close touch with him knew how big Cody's heart really was.'

His strength was a sight to see as he tightened bolts and pulled wires taut with his hands as well as any man using tools, but there was one small thing about him not many people knew. He was always chewing and, being a strong type of man with a big voice and big ideas, people thought it was tobacco but it was only chewing gum.

We missed Cody. Cody was the first civilian to be buried at Aldershot. It was the biggest funeral ever known for a long time for an ordinary man with over 50,000 people lining the 2½-mile route from his home in Ash Vale to the Military Cemetery. Highest ranking officers of the War Office and Admiralty, and Members of Parliament were there to pay their respects. Army, Navy and Royal Flying Corps personnel were represented in large numbers, the pipers of the Black Watch headed the cortege, the Royal Engineers formed a guard of honour and hundreds of wreaths were placed by the graveside. Lela's wreath was in the shape of Cody's steering wheel, the Brooklands' wreath a four-bladed propeller, the Royal Aero Club's a broken column. There never had been a more impressive funeral at Aldershot. I paid a brief visit to Lela to offer my condolences and my assistance in any way. Several funds were opened for her. One pilot, Salmet, collected £25 by charging for his autograph while touring the north of England and we at Larkhill raised £71. With Cody gone, the death toll was now at thirty-seven. Not long after Cody's death, the unhappy news that Vlaicu, the Roumanian pilot I'd met, had been killed when his machine also broke up midair. He was about as well-known in Roumania as Cody was to us in Britain.

Cody's flying skills took him from victory to victory. He'd taken the 1910 and 1911 Michelin Cups, he was the only competitor on a British machine to complete the 1911 Round Britain Race, and his greatest success, the winner of the 1912 Military Trials. I believe his *Cathedrals*, though big, were easy to fly but how I wished he'd changed his ideas about design. The tragedy of it all was that the Round Britain Seaplane race was to be his last competitive flight. Backed by his hard-earned capital, he'd only just formed a company and might have soon retired from flying altogether to live a less hazardous life.

The Round Britain Seaplane Race, a passenger-carrying contest would now commence without its most interesting contestant. Bristol had not entered since the company was concentrating on manufacture and sales, although Coanda had already designed a seaplane, its prototype being tested by Busteed. Burney too, was still experimenting with his hydroplane, the one I'd tried out at Milford Haven. Great strides had been made in the development of the land plane, but as for seaplanes the idea had only just been taken up seriously in Britain, and some thought the contest came a year too soon. France was so much more advanced than we were and had already raced seaplanes. To encourage the growth of the seaplane in Britain, the *Daily Mail* was offering two prizes for contests over water, one for the Round Britain Seaplane Race of £5,000 and the other, a Trans-Atlantic Contest of £10,000 covering the sea-crossing of approximately 1,880 miles was Lord Northcliffe's third great £10,000 prize promoting aviation. As to be expected letters of protest came pouring in, *An inducement to suicide... It should not be allowed...* More immediate was the Round Britain Seaplane Race, set for 16 August 1913 to be completed in 72 hours, total distance 1,540 miles, but everything about this race was to be all British, British subjects, British planes. As in the 1911 *Daily Mail* Round Britain Land Race, five parts on the plane and five on the engine were to be sealed, and the route was to be divided into stages, Southampton, Ramsgate, Yarmouth, Scarborough, Aberdeen, Cromarty, Oban, Dublin, Falmouth, Southampton, with alightings made on water only.

There had been only four probable contestants – Cody and Evans, Radley and Gordon England, McClean and another, Hawker and Kauper, but the race turned out to be a bit of a washout. The Radley-England machine for the contest was withdrawn. McClean withdrew and with the death of Cody, only the Sopwith team, Hawker and Kauper was left, then the 16 August came. Hawker and Kauper made their first attempt but it ended at Yarmouth when Hawker collapsed, according to diagnosis, of sunstroke and fumes from the engine. They made their second attempt from Southampton and flew well as they followed the coast of Britain, reaching Scotland without serious mishap after flying about 500 miles in 10 hours, very good going. They continued the next day, but Hawker had some engine trouble just a few miles short of Dublin, side-slipped and landed in the water. He escaped injury but Kauper suffered a broken arm and had to spend a short time in a Dublin hospital. Though they could not continue they did not go unrewarded. The *Daily Mail* gave Hawker a personal gift of £1,000 and Shell presented him with a silver scale model of the Sopwith Seaplane, made by Mappin and Webb and, Hawker and Kauper being Australians, Sir George Reid, High Commissioner for Australia, sent a cheerful wire to them which read, '*Win or lose, Australia is proud of you*'. The *Daily Mail* proposed holding another £5,000 Round Britain Seaplane Race the following year, for which many more contestants were anticipated.

A bet of £500 was made between Handley Page and Pemberton Billing which aroused considerable interest at Brooklands. Billing, brother to Eardley Billing of the Blue Bird Café, had been building machines as early as 1908 without much success. He was well-known in the yachting world but he could not fly, neither could Handley Page. The bet arose from a friendly argument when Handley Page was telling Billing about his new biplane reputed to be very stable. '*The Page machine is so stable that anyone could learn to fly it in twenty-four hours,*' he stated, cryptically adding. '*Anyone with enough sense could learn to fly on any machine in twenty-four hours!*' and bet Billing that he couldn't fly on one of his own machines or any other in that time. The bet was taken up, and finally agreed upon that the first of them to fly in the quickest time would be given £500 by the other. That was a lot of money, enough to buy an aeroplane, so with £500 at stake, Billing wasted no time. Early next morning, 17 September 1913, Billing was at Brooklands determined to win the bet. It was raining, but a number of people who'd heard about the wager turned up to watch him roll out Ducrocq's old Farman. He was ready to fly; it was 6am. Robert Barnwell, Chief Pilot of Vickers, had been persuaded to help, and after verbal instruction and a few short flights with him, Billing went up solo. Poor Billing performed some horrible actions, but fortunately the Farman was a safe machine and he pressed on regardless, but he did not realise he was running short of petrol until he saw people frantically waving petrol cans from the ground. After two hours of intense flying, he felt he'd learnt all there was to know and landed. An Aero Club official had been difficult to find at short notice, but someone arrived and by sheer luck he passed. Pemberton Billing became a qualified pilot in just about four hours and the licence he proudly possessed, which he could wave under Page's nose, was No 632. Page intended to make his try at Hendon, but retired from the contest on hearing Billing's achievement and paid up the £500.

During the same month, September 1913, news reached England that the Frenchman, Adolphe Pégoud, was doing all kinds of stunts in the air such as deliberately flying upside down and looping the loop. *What kind of flying was this?* Was it really possible? First of all, Pégoud had strengthened his machine, a Blériot, to stand up to the strain, then had prepared himself by spending periods of fifteen minutes strapped in the Blériot as it lay upside down on trestles, and then gave public displays at Buc aerodrome much to the delight of the French who looked upon him as a national hero. *Monsieur Pégoud, the Inventor of Looping*. The Brooklands management, quick off the mark, invited him over to show us what he could do. Many did not believe his claims and some were immediately against the idea. *Freak flying... mere sensationalism,* but Pégoud came over to us with Blériot on 25 September 1913 for a three-day demonstration. We were very lucky to get him to Brooklands so soon, and people from all over the country, including most of us at Larkhill, went to Brooklands to watch.

He did not look like a daring flyer. Amid great excitement, Pégoud got into the Blériot, was well strapped in, and took off, climbing to about 4,000ft. Field glasses were focused on him as he began his fascinating display, '*Remarkable... What a man!*' The crowd went wild. '*I've never seen a bird fly upside down like that... The most wonderful thing that's ever been done in the air...!*' He did four distinct feats, an 'S' shape starting from a dive, a 'Z' shape promoting a tail slide and repeating the motion, a loop by diving down and finishing in a circle, and flying upside-down and then rolling over. He'd done all he said he would. *Flight* summed up the general feeling:

> 'When we first heard of Pégoud and his upside down flights, we not unnaturally regarded it more in the light of a sensational stunt than anything else. Having seen his actual performance, we are of the opinion that it is the most scientific Flying Exhibition that has ever been made.'

C.G. Grey referred to the idea of looping as lowering the status of an aviator to that of a music-hall artist or circus performer, and had received a long, indignant letter from an offended Harry Houdini, the famous 'Handcuff King', who had been the first to fly in 1909 in Australia. His letter was reproduced in *The Aeroplane*, a paragraph of which read:

> 'I wish to protest against the manner in which you wrote several of your articles in which you seem to hold very low opinions of music-hall performers. Whenever you speak of trick flying or Pégoud, you say that they need not degrade themselves to the level of music-hall performers. Twice you have published this, and I think you are either ill informed about music-hall performance, or believe that a Pilot is a Superior Person.'

Pégoud is credited by everyone as the first person to loop, but Russia claimed that they'd been the first, saying that a Russian lieutenant had looped in a Nieuport some time before Pégoud, and adding this officer had been penalised to thirty days detention for his behaviour. I certainly do not believe the Russian claims since Russia was still a backward country and, eager to show herself in the aviation world, made a habit of claiming to be the first in several things which had so often proved invalid.

Now we had seen it, it was only natural that many pilots in Britain would take up Pégoud's specialised type of flying, and the first one to do so was B.C. Hucks whom I knew very well since the Harvard days. Soon after seeing Pégoud at Brooklands, Hucks went to Buc to train with Blériot, then Grahame-White arranged for him to give a public demonstration at Hendon on 27 November 1913. Great crowds gathered to see him, then during his display, Gustav Hamel became the second Englishman to loop, but not before he'd suffered some anxious moments when he dived, turned his machine upwards and could not bring it over. On becoming more proficient, Hamel received a Royal Command to give an exhibition before the King and Queen at Windsor Castle, while Hucks toured the country giving displays and making a fortune. Meanwhile, Pégoud in great demand toured Europe, and besides doing the aerobatics we'd seen, he was now cart-wheeling in the air, flying upside down without a safety belt, and climbing from one wing to the other of his Blériot as it floated earthwards, regaining control before it reached the ground. Then came news of J.B. Thornly, son of Professor Thornly of Merton Hall, Cambridge, with the latest invention for looping, wheels above as well as below in case of an inverted landing. While this new fashion of looping was the craze, Captain Bertram Dickson died quietly at Lochrosque Castle, Ross-shire, after a long three years of suffering and stomach pumps since his air collision. He was a very charming, aristocratic man who had a passion for flying and, in 1910 flying his Farman, was the first to break a world record for Britain, a duration flight with a passenger.

Many famous French pilots had left the flying scene. They included Santos-Dumont, the Brazilian living in France, retired from aviation in 1910, Blériot had given up active flying, Louis Paulhan had decided to spend his time on a horticultural project on the Riviera, and Lieutenant de Conneau, known as Beaumont, had also turned his mind to other things, Roger Sommer of France with 180 machines to his name decided to leave aviation to others, as did Robert Esnault-Pelterie of the REP machines and engines. Several well-known French aviators had died too, including Charles Voisin in a motor accident, and the Nieuport brothers of the popular Nieuport aircraft, who'd both died in air crashes. As for the Deperdussin Aircraft Company, its founder was arrested for colossal frauds and the British Deperdussin firm went into liquidation. The charge against him was that he'd obtained, by fraud, over a million pounds from a French bank by bogus business documents. He owned two aeroplane factories and the Champagne Aerodrome and, in recognition of his services in aviation, had been made a Knight of the Legion of Honour. He was also connected with theatrical enterprises, a private sanatorium and a hot-air Therapeutic Institute, all build up from funds obtained by fraud. When he was eventually brought to court, he pleaded guilty to all charges and was finally let off under the First Offender's Act. His bank could claim nothing.

Early October 1913, I returned to Brooklands having been at Larkhill over two years, and found some digs at 3 Kings Road, Walton-on-Thames, nearly three miles away. The War Office wanted to take over Larkhill as the ground was required for Artillery Ranges, and we were expected to be out by about March 1914, so this meant we had to make various adjustments and do what we could from Brooklands. The only active schools at Brooklands nowadays were Bristol and Vickers although the Ducrocq School was training the occasional pupil or two. Avro had not trained a man for over a year as the company, now very much more established, was kept busy meeting orders for the Government. Among the many new men were Skene and Vincent Waterfall flying for Martin & Handasyde, and the Sunbeam Engine Company had acquired a place at the ground. Alcock was known as 'the Sunbeam Man' as he flew a Maurice Farman fitted with one of their engines.

Merriam was the Bristol instructor at Brooklands, a most painstaking man and the finest instructor I ever knew. He'd not been flying for very long, having qualified with us in February 1912, Licence No 179, but he had the enviable record of teaching more men to fly than anyone else. About seventy pupils, maybe more had been put through their tests by him during the short period he'd been with the company. Much of his success was due to the pupils liking his methods of teaching, and all his instruction had been given on Box Kites, a machine he loved and which gave many a young man confidence in flying under his expert tuition. He never hurried anyone. The nervous ones quickly overcame their difficulties when they realised they could take their time. He knew how it felt. He used to get giddy with heights and thought he would never live through his first flight. Shortly afterwards he went to Larkhill for a spell and was involved in an accident in which a pupil died and he escaped injury. They'd been flying a Bristol monoplane with dual controls. Perhaps for a few seconds neither had been in control due to misunderstanding for the machine banked, and dived to the ground. It came as a terrible shock to Merriam since he'd been free from serious accidents, and for a long time he was visibly upset.

We were nearing the end of 1913 and the aeroplane had become quite settled in its design now, but there was one we were especially curious about, one we'd never seen, in a shed at Brooklands. There was an air of mystery and secrecy surrounding it and a hefty guard stood around, even when the place was locked up. All we knew was that inside the hangar there was a machine belonging to Prince Serge de Bolotoff, a young Russian nobleman living in England and elder son of Princess Wiasemsky, and that several influential Englishmen including Lord Marcus, were on his board of directors. We were naturally curious when we heard it was to make its debut, and gathered around for the great moment. It turned out to be a huge triplane named after its owner. We watched and waited anticipating great things as the engine was revved up, but nothing happened so its enthusiasts gave it an encouraging push which did the trick. It lurched forward. At least it was a start, but then it moved around jerkily without a sense of direction and fell over to one side, it hadn't the slightest chance of ever flying.

A more interesting machine came to light called the Cedric Lee or Lee Richards, an original patent by Kitchen of Bowness-on-Windermere bought by Cedric Lee and Tilghman Richards. Archibald Sinclair financed its development but it had been kept a secret for three years. What was unusual about it was its shape as the wings, unlike those of a conventional plane, formed a circle and it was nicknamed *the Doughnut* for it looked like one being completely round with a hole in the middle. The purpose of its circular shape was to achieve automatic stability, but from the start it was very much out of balance and Gordon England, who'd helped with its construction at Shoreham Aerodrome said, regarding its early flights, 'It's powerful. On one of the test flights I put the throttle back and it shot up, looped and landed on its back.' With some modifications it might have proved itself the wonder machine of the day, but it never amounted to much... a small fortune had been spent on its development.

Around this time a new Personal Accident Insurance came out for pilots, good news as no company previously would insure us, too great a risk. It was instituted by Holmwood and Holmwood of Gracechurch Street, London, and C.G. Grey was one of the first to report the news. 'In consideration of an Annual Premium of £21 the firm undertakes to pay £1,000 for the loss of both hands or feet or loss of sight of both eyes, £500 for the loss of one hand or one foot or loss of sight of one eye, or £5 per week during temporary total disablement, payment being limited to twenty-six weeks, excess of first two weeks.' A step in the right direction.

My son, Jack, was born in October 1913 and C.G. Grey, once more in a teasing mood, advertised the fact by reproducing in his magazine a photograph of my wife and I seated in a Bristol, but under which was written, and this only, 'Isaiah IX, Verse 6, Lines 1 and 2'.

Bristol had grown from strength to strength and for the third year running exhibited their latest machine at the Paris Aero Show held at the end of the year. They'd displayed the Prier monoplane in 1911, the Coanda monoplane in 1912 and their 1913 star exhibit, the Coanda two-seater biplane fitted with a new bomb-dropping device, an invention by Coanda, containing twelve bombs which could be released by a hand lever in the observer's seat. It was obvious that aircraft were being prepared, thus equipped, for hostile action in case of war, first the Vicker's Gun Bus and now the Bristol bomber.

One day in November I called upon a mechanic to make minor alterations to the tail-piece of the new Coanda biplane which I was to fly at Brooklands. As he tilted up the tail wings according to my instructions, I was called to the telephone. It was Coanda at the other end sounding very indignant. Some kind person had phoned the office to tell them about the alterations I was doing and I am sorry to say Coanda and I quarrelled. *'You must not alter the machine in the slightest degree!'* I told him it was tail heavy and I intend putting it right. He told me that I was not in a position to carry out structural alterations, the calculations were correct. I repeated it was tail heavy and as I have to fly it, I want it right, *'It's my life at stake, not yours! It feels like a flying flea!'* and he repeated, *'No alterations!'* I'd seen enough deaths and I certainly wasn't going to fly a machine that was tail heavy and unsafe, I didn't trust it. Stanley White was the next to speak with me on the phone. 'Coanda's instructions must be carried out,' I was not to make any changes to the aeroplane! I explained that if this alteration was not done I would not fly it. There was silence at the other end as Stanley White apparently discussed the matter further with Coanda. *'No alterations will be carried out!'* I promptly said, *'If you are not satisfied this requires attention and believe that I should fly the machine as it is, then I have no alternative but to leave your services!'*

I was very happy in Bristol's employment, there was a friendly home-from-home atmosphere with them and they were very good to their pilots. For example when they lost a pilot, they did all they could for the bereaved family taking a personal interest. I was well established with them, received good remuneration, £350 annually and a third of prize money, and was very sorry that this incident had created an unpleasant situation but, most of all, it was regrettable that Coanda had been involved. I could never forget the wonderful hospitality he and his father had shown me in Roumania. The incident was not settled and most reluctantly, but very positively, I left. I was in the wrong. I'd been too hasty. Coanda was a very clever designer, first making the very successful Coanda monoplanes for the company then, using the same fuselage, the Coanda biplanes which were much better but, according to my judgement, a bit tail heavy, a judgement which was later proven correct when the overlooked defect became evident but it came too late for me.

I had found a new position testing planes for Tommy Sopwith.

CHAPTER TEN

With Sopwith Flying the Tabloid Biplane

I knew Tom Sopwith well, a very clever young man. He came from quite a wealthy background and as a result of the great sums of money he'd won in flying and other racing pursuits, he formed the Sopwith Aviation Company to expound his ideas of aircraft design. Although Sopwith was an exceptional pilot, a credit to British aviation while touring America in 1911, he was not flying any more and for a while had run a flying school at Brooklands with Raynham as his flying instructor, but this he gave up to concentrate on construction having taught only a few people. Raynham went back to Roe. Helping him was Fred Sigrist and as the company expanded Sopwith leased a skating rink in Richmond Road, Kingston-on-Thames, for construction work and put Sigrist in charge. During 1913 the company had got well underway, but as Harry Hawker was Sopwith's only pilot I asked him whether he could do with another since Harry was wanting to return to Australia for a few months. He suggested I came immediately in a temporary capacity, thus enabling Harry to leave whenever he wished. We came to an agreement... the Sopwith policy was to enter all competitions and in view of this I was to receive £5 a week basic rate, £5 for each machine I tested and I would be testing a number, and a share of the prize monies, in all amounting to very good pay.

Little work was now being done at Brooklands, but all new machines were tested there and I would be handling them as they came in. Harry and Tommy Sopwith were roughly the same age, about 26-years-old. I was the veteran being now in my twenty-ninth year. Like me Hawker had been hard up when he entered the flying world, and was one of five Australians, all motor mechanics, who'd travelled to Britain on the same boat in 1911 hoping to learn to fly. A few other men had also left Australia to come to Britain and learnt to fly during this early period of flight. The five, Busteed, Harrison, Stutt, Kauper and Hawker, split up shortly after arrival when three joined Bristol's and all made good. The other two, Hawker and

Kauper, got cheap lodgings in London and worked as motor mechanics until such a time they found positions in aviation. After a year, Hawker was on the verge of returning home as he began to believe he would never find a position in the flying world when Sopwith advertised for a mechanic. Kauper applied, got the job, and told Sopwith about his friend. Sopwith agreed to take Hawker on as an extra mechanic, saw he had the makings of a good pilot and taught him to fly. Hawker received his Licence No 297 in September 1912, and from then on he proved to be an excellent pilot and a great asset to Sopwith.

Sopwith's first original design was a flying boat, the *Bat Boat*. It was the first successful effort to take a boat in the air as distinct from a seaplane which until now was only an aeroplane on floats. On it, Harry Hawker won the Mortimer Singer prize in July 1913. This involved landing both on land and water and Hawker did it from the Solent flying to a point 5 miles inland, alighting and returning to the water six times to take the £500 prize. By doing these land-to-water flights, the *Bat Boat* was also recognised as the first amphibian aeroplane in Britain. Several Naval Air Stations had been recently set up around the coast at Calshot, Isle of Grain, Felixstowe, Firth of Forth and Cromarty, which were being equipped with aeroplanes built to War Office specifications, and one of the first things I did for Sopwith was to deliver the *Bat Boat* from Cowes to Calshot. It was a week before Christmas 1913. With me came Victor Mahl, a new mechanic who was quite a bit older than Sopwith or myself, who was hoping to learn to fly with us. He was excellent in his job, and we got on very well together. When we arrived at Cowes we had the greatest of difficulty in starting the engine because of the length of time the *Bat Boat* had been left unattended, but at last the machine showed some life and we rolled it from the shed and onto the water at the mouth of the River Medina ready for me to test it out. Mahl was returning on his own to the mainland and waving farewell to him, I began moving over the surface of the water. It was wonderful. I swerved in and out of the shipping loving every minute of it until I was in the Solent, then took off feeling completely at ease with the machine as if I'd never flown anything else. Calshot all too quickly came into sight. In front of Naval officers I put it through the tests stipulated by the Navy, and handed it over rather reluctantly for I was sorry to see it go and hoped that all Sopwith's machines would handle as well as his *Bat Boat*.

Another of Sopwith's 1913 machines was the very successful 3-seater Standard tractor biplane which was being produced in numbers for the War Office. They were proving popular with the Service pilots, and my next job was to test a batch of them, part of an order of nine for Farnborough. The first had been flown over by Hawker, but they were coming in from the works quite rapidly and I got them off at a fair pace during the first week or two of January 1914 after preliminary tests at Brooklands. At Farnborough I put them through their Acceptance Tests, which included fast and slow speeds, rate of climb and an hour's duration flight. They were also inspected for quality of materials used, workmanship, and checked to see whether the Government specifications had been adhered to. Mahl sometimes flew with me. It didn't take long to get to Farnborough, a short trip of about 16 miles, but once the engine failed and I was forced to land with my passenger, Lieutenant Mapplebeck this time, coming down close to Lord Northcliffe's home. While we were looking to see what was wrong, Lady Northcliffe appeared and asked us in. Unfortunately we could not accept her kind offer due to lack of time. I still saw C.G. Grey, and told him bits of news for his magazine, and after speaking about this particular delivery and our meeting Lady Northcliffe, he wrote quite a lengthy account of it.

> 'During the past week there has been a considerable amount of activity at and from Brooklands. On Tuesday Mr Pixton flew the sixth of the Army's Sopwith biplanes, 80hp Gnome, to Farnborough with Lieutenant Mapplebeck, RFC, as passenger. To avoid the bad country by Woking, Mr Pixton generally makes this journey by Guildford and on this occasion the engine failed, so that they were compelled to come down close to Lord Northcliffe's place, Sutton Court. Lady Northcliffe kindly invited the Aviators to lunch, but they were reluctantly compelled to decline, and continued their journey to Farnborough as soon as possible. During Thursday afternoon Mr Pixton put the machine through her tests, the high speed reaching 71 miles per hour.'

He continued to report each delivery.

> 'On Friday, Mr Pixton, accompanied by Mr Mahl of the Sopwith Company, took the seventh machine over to Farnborough without incident... On Saturday last Mr Pixton passed the eighth of the Sopwith 80hp biplanes through its military tests at Farnborough, its maximum speed being 74 miles per hour... On Sunday at Brooklands he was flying the ninth and last of this order.'

I tested a few more planes and delivered them to various Air Stations, including the Isle of Grain station at the mouth of the River Thames which was being equipped with aeroplanes to control the approaches to London and was expanding rapidly. The Isle of Grain, as an island, was hardly worthy of its name as one could virtually step over onto it from Kent. Further testing lined up for me included six seaplanes for the Greek Navy, a very large machine with an open fuselage and twin floats, and during March 1914 I tested the first of them over the Hamble River. We were unable to do any seaplane tests on the River Thames near the works as the river authorities would not allow it. Besides, it was far too narrow for extensive tests.

A large *Bat Boat* had been built which Sopwith entered for the 1914 Olympia Show during March, our sole exhibit, but there was another flying boat on display that caused considerable interest, the PB1, the first Supermarine Flying Boat, and the creation of our friend Pemberton Billing. the winner of that famous £500 bet of learning to fly in twenty-four hours. Billing had his works at Woolston, Southampton, part of which we used, and we'd already seen the Supermarine undergoing its trials on the Southampton Waters. Sopwith had a Rolls Royce and he drove down to the Southampton area from time to time taking me with him. One day while talking to Billing I asked how did he come to use the fine name of *Supermarine* for his flying boats, and he explained that submarine refers to underwater, marine on water so why not *Supermarine* above water... boats that fly! While in the area I saw a Model T Ford Motorcar and bought it. Merriam, who had a great love for cars and owned one as early as 1903, was happy to buy my Argyll which had served me so well since 1911.

Hawker's greatest rise to fame came when he flew a small 2-seat Sopwith *Tabloid*. The Tabloid had taken everyone by surprise during November 1913. It was only 20ft long and spanned 25ft 6ins and was drawn up on the works floor at Kingston by Sopwith, Hawker and Sigrist. Hawker had taken it on its test flight flying with a passenger just before I joined the company. It proved very fast, the fastest plane we'd seen and when it was shown to the public it caused a sensation. Everyone was astounded by its speed. Its unusual name came about because of its size but *Tabloid* infringed the Burroughs Wellcome trade name applying to many of their products, so it was made known that the official name of the machine would be the Sopwith Scout. However, the name *Tabloid* stuck and the Chemical Company could do nothing to stop the usage of it as a nickname. So be it. We had a substantial order from the War Office for Tabloids and once completed I would carry out the tests on them.

Meanwhile in Australia Hawker was giving flying displays on his 2-seat Tabloid. Thousands came to Randwick Racecourse, Sydney, to watch him fly, and there he took up the Governor-General, Lord Denman for a flight.

Wherever he went he attracted crowds. During his absence, a man was impersonating him, but he didn't last long. *Flight* got hold of the story.

> 'When Mr Hawker returns to this country from Australia, he will have to search out Mr Rutherford of Gainsborough and have a few words with him. Mr Rutherford rode up to one of the hotels in Lincoln the other day on a motorcycle and said he was Hawker, come to do some Exhibition Flying in the neighbourhood. He was of course immediately the hero of the place and was introduced to all and sundry. He had great tales to tell of his flight Round Great Britain, and what he was going to do in the future. So friendly did he become that he promised to take many of the guests for a joyride so soon as his mechanics should arrive with his machine. Of course, there is always somebody who can't leave a poor chap alone, and the killjoy in this case was a Policeman who turned up and arrested the giddy Rutherford for stealing the motorbike which he had hired at Gainsborough and had forgotten to return. The Police found that he was in possession of nine-pence, which goes to prove the old saying that it is not always necessary to possess money to be happy. Incidentally, he said he had had over a thousand offers of marriage. Perhaps when he finished his four months imprisonment he will consider one of them.'

While I waited for the Tabloids to appear, a Sopwith *Tweenie* 2-seater dual control sociable biplane built for the Admiralty arrived from the works, a slightly larger version of the Tabloid but smaller than the 3-seater. I tested it and found it flew right away without the need of any adjustments and a few days later handed it over to Lieutenant Spencer Grey of the Royal Navy, Licence No 117, who piloted it from Brooklands to Hendon, but not without considerable difficulty. It took him two and a half hours to do the 20-mile journey, coming down four times because of mist and rain. Next day, the Right Honourable Winston Churchill, First Lord of the Admiralty, went up on *Tweenie* with him and took the controls. He'd taken a great personal interest in the development of the Naval Air Service. At last a Tabloid came in from Kingston during the second week in April, and whereas Hawker's Tabloid was a 2-seater, this was a single-seater, the first of a batch for the Army for scouting purposes and capable of carrying more fuel. With sheer delight I flew it. It was like handling a racehorse all other machines being carthorses by comparison. I loved the ease of the controls, the extreme stability and quick recovery from dives, it was difficult to make it side slip, and *The Aeroplane* reported:

'Mr Pixton made his first flight on the newest Sopwith Tabloid, it was what the French would call an *emotioning performance* as he had never flown one of these small projectiles before, but he landed safely at about twenty miles an hour.'

It was the first machine to really roar with speed, but there was a snag. So great was the general respect for the little machine that I found I was being handicapped out of first place in races. I'd had the machine for only a few days when we entered it for the £50 Easter Aeroplane Handicap at Brooklands comprising of a short course of 9 miles to Coxes Lock Mills, there and back twice from a standing start and a flying finish. A huge crowd was present, and the first to go in a field of eight was Merriam on a Bristol Box Kite. I was at scratch and last to leave. I raced full out and overhauled the early starters but didn't quite make it. I did not have sufficient time over the short course to pass the two leading machines, the Blériot and Vickers which were so much slower than the Tabloid. Another few seconds and I would have been the winner. Barnwell was first, Knight second, Self third, Merriam fourth. It was ridiculous, to be nearly beaten by Merriam on a Box Kite! If I were constantly handicapped like this I might never win a race, and I mildly complained that the handicap was too stiff for such a short distance, that six and eight minute allowances for a nine mile course is hardly sensible. Hawker would say the difference between a tractor biplane and a Box Kite was like comparing a car with a bullock wagon.

\multicolumn{4}{c}{**Easter 1914**}			
\multicolumn{4}{c}{**Brooklands Handicap Race**}			
Pilot	**Plane**	**Minutes Lead**	**Results**
MERRIAM	Bristol Box Kite	8.24	**4th**
KNIGHT	Vickers	6.42	**2nd**
BARNWELL	Blériot	4.12	**1st**
ELSDON	Vickers	3.12	
ROEMPLER	DWF	2.20	
ALCOCK	Maurice Farman	2.6	
WATERFALL	Martinsyde	0.30	
PIXTON	**Sopwith Tabloid**	**Scratch**	**3rd**

Afterwards we gave exhibitions of flying for the benefit of the spectators. One thing I did on the Tabloid which the public enjoyed was to charge down at full throttle, shut off the engine and go into a steep upward glide, judge by instinct when stalling point was near, then flatten out and repeat the performance. At one time Busteed, on the new Bristol Scout, and I were in the air together giving a display which was described as, '*A trifle too thrilling*,' followed by, 'There will have to be some definite and rigid rules of flying if these ultra-fast machines are to be used in Aerodromes without causing fatal accidents.' Robert Barnwell thrilled the crowds by looping the Tabloid and became the first Brooklands man to perform the act. Up to now most looping had been done by Hendon pilots. On landing after a display Barnwell declared, 'I was beginning to wonder where the world had got to.'

I still had no desire to attempt stunts in the air, not even on the Tabloid.

CHAPTER ELEVEN

With Sopwith Winning the Schneider Trophy

Suddenly Sopwith decided to enter me for the annual Schneider Trophy International Seaplane Race. It was being held near the end of April at Monaco and already we were in April. Will we have time? Sopwith often left things to the last minute but once he had set his mind on something he saw it through regardless of the odds. *We needed a Seaplane!* A snap decision to adapt the Tabloid was made. From that moment on everything was rushed, the Tabloid was stripped of its wheels, a single control float put in its place, and its 80-horsepower Gnome changed for a newly marketed and untried 100-horsepower single valve nine cylinders Gnome, the lightest engine in existence. We'd first heard of Jacques Schneider during December 1912, when his splendid offer for the promotion of seaplanes was made known at the Gordon Bennett Banquet of the Aero Club of France. 'Monsieur Schneider has offered for International Competition a trophy of the value of £1,000 to go to the club which the winning pilot represents and in connection with this, he also offers to give £1,000 annually for three consecutive years.' The rules binding the Schneider competition came under the FAI, Federation Aeronautic Internationale. All great races were run under rules formed by them and the British Aero Club had been a member of the International body for a long time, since its early days as a Ballooning Club.

The Schneider Race could go on for many years as a country had to win three times in five consecutive contests to obtain the trophy, and each year the winning country would have the honour of organising the next competition until eventually the trophy was claimed. Basically, it stipulated that the course be at least 150 nautical miles, over sea, and point-to-point, circular or angular. Apart from the £1,000 prize for first place in the first three years of running, secondary prizes covered by the entry fees would be awarded, and for the fourth year onwards awards would then be covered entirely by entry fees as by that time the contest would be holding its own. It was now in its second year. The first race, held in 1913 in France in conjunction with the Monaco Seaplane Meeting, was described by C.G. Grey as, *'something of a fiasco'*. Prevost, the Frenchman whom we came to know well during the Military Trials, had been the only competitor out of four to finish. He believed he'd won at an average speed of 61 miles per hour, but he did not pass the finishing line and was called upon to do so after a delay of nearly an hour with the result that fifty-eight minutes was added to his time of 2 hours 50 minutes. Because of this, his speed was officially recorded as 45 miles per hour. The confusion arose when a gun fired as he came in and, believing he'd won, Prevost had put away his machine.

The regulations for the 1914 Schneider, translated from the French, were as follows:

'The distance to be accomplished for the Jacques Schneider Maritime Aviation Competition in 1914 is 150 nautical miles. The Competition will take place exclusively at sea on a closed circuit having a minimum course of five nautical miles and situated outside any port or closed harbour. Alightings are permitted. Competitors may start at any time they wish between 8am and the official hour of sunset. Only one attempt is allowed, and before starting a competitor must notify the officials of his intention to take part in the race. Two copies of the final regulations will be handed to all contestants, and one copy of these regulations, signed and approved, must be handed by the competitor to the official before starting the competition. The competitor will then navigate his machine over the line of departure, rise, and make a tour of the course with at least two alightings on the water at points indicated by the officials. This having been accomplished, he must proceed without alighting to make the first lap of the circuit, clearing the line of departure in full flight. He must then continue the course until the whole distance has been completed, when the arriving line must be passed in full flight and the alighting made on the course. The race will take place at Monaco on 20 April 1914.'

After the haste of preparing the converted Schneider machine we rushed off to try it out on the Hamble River. Sopwith, Mahl and others were watching from the bank as I taxied out on the Tabloid now sitting high on its wide, single float, but as soon as I opened the engine the thrust made the float dig down in the water, and I shot head over heels out of the cockpit and into the water. The water was very cold. I felt it slowly seeping through my clothes, an unpleasant sensation, but out of the corner of my eye as the machine toppled over I'd seen Mahl without a moment's hesitation, jump fully clothed into the water to assist me. Although I naturally appreciated his action immensely, I told him as we swam back together, that it was a foolish thing to do, jumping into icy cold water like that, I was perfectly all right and prepared for it. We didn't have a boat at hand to tow the machine to the riverbank so we left it as it was, nose down in the water, until the tide went out but it drifted down the Hamble and into the Southampton Water and lay there for a day. Our men had a most difficult job in getting it out but finally, with a rope fastened around its tail, it was pulled ashore on the change of tide in the early hours of the morning. As I looked at it covered with mud, I felt the dip in sea water might not have improved the new engine. We hauled the Tabloid onto a vehicle and rushed back to the Kingston works where the single float, which had been the cause of all the trouble, was cut in half and fitted onto a new frame to provide two floats for better stability.

With this done and the Tabloid had dried out a bit, we took it for further trials, this time to the Thames near the Kingston works as we had only a few days before it had to be packed for Monaco. If we didn't make much headway with the trials we would miss the deadline for entry. I was ready to try it out for the second time. The floating tests proved just about passable although I would have liked the Tabloid sitting more proudly on the water with the floats not so submerged, but just as I was about to take off, officials appeared on the scene and were most unsympathetic as to what was going on. They were the men from the Thames Conservancy, *'You can't fly here!'* It was no good trying to argue with them. They absolutely forbade us to fly over the Thames, so we had to dismantle the machine and go elsewhere. *Where?* We rushed down to Richmond early next morning just below Teddington Lock and put it on the water near Glovers Isle, which came under the Port of London Authority. We received permission to fly. With time running out quickly, I made another attempt to test it. Once satisfied with its control on water, I flew off with the engine misfiring a bit to Eel Pie Island, a little island in the middle of the Thames at Twickenham, and managed to reach a good speed of 85 miles per hour, but it was barely a 3-mile flight, just a hop in fact, there just wasn't the time to fly any further. All this, the conversion of a landplane to a seaplane, the tests at Hamble and on the Thames, had been done in four days. Immediately after this last flight the machine, though inadequately tested, was dispatched at once to Monaco, and without a moment's delay we also left.

C.G. Grey put a full-page picture of the Schneider machine, taken on the bank near Teddington Lock, on the *The Aeroplane* cover with these two words, 'Good Luck' and reporting, 'Mr Sopwith does not expect to win, he merely hopes that his machine will put up a respectable performance.' This was so. Sopwith did not expect to win.

On arriving in Monaco we made our way to the Hotel Bristol in Monte Carlo, overlooking the Bay of Monaco where we had been invited to stay. The Schneider course, marked out with buoys, lay directly in front of the famous Monte Carlo Casino and the pigeon shooting ground, the *Tir-au-Pigeons*. We were just above the Casino, and had a fine view of where the field of action would be, but some very anxious moments were spent wondering whether our machine would show up on time since we'd known instances when machines had not arrived to schedule. Much to our relief, ours appeared on Thursday, four days before the race. Mahl wasted no time in getting it out of the packing case and, to our eyes, it looked splendid on the floats but luck was against me regarding further tests. A mistral suddenly blew in from the Sahara and it looked very doubtful that I would be able to put it under any kind of test before the racing. These mistrals could die down as suddenly as they arose, but even if the race were postponed it would probably be run at the first opportunity still leaving me without much time for test flights.

Quite a bit of seaplane racing had been done at Monaco over the last year or two. In fact it was the first major centre in the world for the event and had seen many a water plane go down or overturn. It was becoming so well known for flying organised by the International Sporting Club of Monaco that many British visitors were already in Monte Carlo on our arrival having come to see the Monaco Meeting. The meeting, which included in its programme the Monaco Aeroplane Rally and the Schneider Race, had opened at the beginning of April. The rally was all over and done with, which had consisted of seven routes, each of 1,293 kilometres from Paris, London, Brussels, Gotha, Madrid, Milan and Vienna, all merging at Monaco. And Garros, who was the winner, was entering the Schneider.

Garros was the favourite to win!

Representatives from four countries were entering the race: France, America, Switzerland and Britain. Germany had entered but would not be competing, having crashed both their machines before the race. The Tabloid was the only British machine present. Three of the entries were biplanes, the Curtiss, the FBA flying boat and the Sopwith, but the French favoured monoplanes as they thought biplanes were not good racers and believed that it was not possible to build a fast machine with engine power below 160-horsepower. All their machines were fitted with one. Certainly the French machines, the Nieuports and the Morane-Saulnier, were considered by all to be the most sophisticated machines in the race. The reason why Prévost, last year's winner, was not entering was due to bad luck during the French eliminating trials as eight machines were entered, but only three could participate, thereby making it necessary for eliminating trials which were held over a 10-kilometre course, four times around, and only Espanet and Levasseur did the four laps at an average speed of about sixty miles per hour. Garros did a single lap which was more than the remaining others did and these three were officially named for the Schneider. Prevost, Janior and Brindejonc des Moulinais were in reserve. In the little harbour round a point in the bay allotted to competitors' machines, our little Tabloid looked so slight and fragile beside its big competitors that no one was taking it seriously. Even so, Mahl insisted on sleeping close by to guard it until the race was over instead of enjoying the comfort of hotel accommodation.

20th April 1914

THE SCHNEIDER TROPHY SEAPLANE RACE

Pilot	Nationality	Plane	Engine Type
GARROS	**France**	**Morane-Saulnier monoplane**	**160 hp Gnome**
ESPANET	France	Nieuport monoplane	160 hp Gnome
LEVASSEUR	France	Nieuport monoplane	160 hp Gnome
WEYMANN	America	Nieuport monoplane	160 hp Le Rhone
THAW	America	Curtiss biplane	100 hp Curtiss
BURRI	Switzerland	Franco-British Aviation biplane	100 hp monosoupape Gnome
CARBERY	Britain	Deperdussin monoplane	160 hp Le Rhone
PIXTON	**Britain**	**Sopwith Tabloid floatplane**	**100 hp monosoupape Gnome**
STOEFFLER	Germany	Aviatik. Crashed before race	
FOKKER	Germany	Fokker W1 Crashed before race	

On Friday we were invited to a dinner at the Hotel de Paris by Monsieur Jacobs, Vice President of the Federation Aeronautique Internationale, held in honour of the Monaco Rally competitors and the Schneider Trophy contestants. It proved a grand opportunity of meeting well-known French pilots and getting on intimate terms with our competitors. Prévost was sorry he was not flying this time. He was an excellent pilot and holder of World Speed records, but even without him we had stiff competition with Garros and Espanet, both of whom had entered the 1913 Schneider, also our friend Weymann, the American whom I'd got to know during the Round Britain Race of 1911, knew the course having almost completed it last year. So I would be racing against competitors who had contested the 1913 Schneider, Garros, Espanet and Weymann. Weymann, recently on the board of directors of Société Nieuport and well used to flying Nieuports, had won the Gordon Bennett race in 1911. During dinner we fell into conversation, and he pointed out Garros. Of course, I knew of Garros's reputation but had never met him.

There he was, France's most popular pilot and one of the world's finest who was expected to win the Schneider hands down!

Garros had crossed the Mediterranean in September 1913, 453-miles non-stop in eight hours, and the Council of the Legion of Honour conferred the Chevalier's decoration upon him for it. Not only that, the French Academy of Sports had voted him the Best Performer of the Year 1913 for his flying activities. Aviation was still being considered a sport, its pilots sporty types. Lieutenant Conneau, otherwise Beaumont, had won the honour the previous year, and was managing the Franco-British Aviation Company. Its FBA flying boat that Burri was racing, was designed and built by him with the help of another French pilot. He was keeping it very secret, no one could get near it. My British rival, Lord John Carbery, who had recently obtained a French Flying Licence, was flying for pleasure usually under the name of Carbery, and for the Schneider had borrowed a Deperdussin having crashed his own machine. We knew him well for he flew at Brooklands and Hendon from time to time. I wished him luck.

With only one day to go I was up early on Sunday eager to do a test flight as I'd not touched our Tabloid since the short flight to Eel Pie Island. I was pleased to find the mistral had ceased. No one was about. We got the machine out and I took off, it rose smartly and I tested it for sharp banked turns as the course had four angles, one being most acute, which had to be taken 28 times. It felt fine and after ten minutes, not wanting to display it too much, I put it down into the harbour again. Several little jobs had to be attended to and Mahl got to work on them. The stay wires attached to the floats bent under the strain of the flight and heavier ones were necessary. We changed the propeller, and as we did not know the fuel consumption of the new monosoupape engine, being virtually untried, a supplementary petrol tank holding about six gallons, was lashed in the cockpit and connected to the main tank. I was to carry 30 gallons, enough we believed, to last me the race without coming in to refuel. 'Sopwith' was boldly written on the fuselage, and our competition number '3', was on the rudder fin. All in all it looked very handsome, but we were still being ignored by the other pilots. I did not make another flight this day but trusted all would go well when I would fly it around the difficult four-point circuit the next day.

Monday morning came – *the day of the race!* A crowd was gathering in the bay and on the slopes of Monte Carlo to watch the proceedings as I did a test flight just to check that all was well. A moderate easterly wind was blowing and in front of us lay twenty-eight laps of the 10-kilometre course, an equivalent of 172 miles. As stated in the rules we could start at anytime from 8am onwards, but I'd not decided when I would go, I would watch the others to see how they fared before setting off. On the dot of 8am the starting gun was fired. Levasseur was away, then Espanet in their Nieuports, and almost immediately after them, Burri. The Nieuports were heavier and larger than the Tabloid which was compact and of wooden construction, and I noted the French take-offs were sluggish, also they laboured the touch downs during the first lap when those two landings on the water had to be performed. All three of them looked slow in the air, they were taking flat turns at the points and flying very high, it looked as though they were after an altitude record. I was astounded by their technique. *I knew then, I had a very good chance of winning* and decided to join in at once.

Mahl tuned up the engine and just fifteen minutes after the first man I was in the air. The compulsory landings in the first lap did not slow me up as I touched down and skimmed across the water far faster than my three rivals, making the two touches unmistakable, and at each pylon I cut my turns and banked well over much to the surprise of everyone. But I had a special technique. Most pilots flew close in to the pylons and then went wide, but the quicker way was to advance fairly widely then skim round. It was at once seen that the Sopwith was very much faster than any of the others and was reported that no waterplane had ever been seen banking like that at Monaco, if anywhere, that I gave full warp and banked from 60 to 70 degrees every time. Lord Carbery followed soon afterwards on his Deperdussin, making five of us in the air at one time, but he gave up after one lap. Each lap was roughly 6 miles and it was taking me approximately four minutes each time around which meant I was flying at a handsome speed of about 90 miles per hour, and I felt I could keep this up throughout the race. Again the reporters were busy:

> 'Pixton lapping with the regularity of a cheap alarm clock, if not with that of a good watch, his lap times being after this fashion of a series, 4 minutes 5 seconds, 4.08, 4.04, 4.06, 4.09, 4.09 and so on.'

As no signals were given from the ground, I had my own way of counting the laps having pinned bits of paper about the size of postage stamps onto the dashboard which I'd number from 1 to 28. At the end of each lap I tore one off. 'The little biplane went gaily on,' they said. I knew I was doing well.

> 'For an hour or so it was simply a matter of admiring Pixton's turns and figuring out how much faster his little Sopwith was than either of the other machines, for he had matters absolutely in his own hands as regards speed from the very start.'

Such was the amazement of Garros, who was still on the ground. *'I might as well not even start as I could not possibly beat Pixton.'*

I continued without a hitch until the fifteenth round when the engine suddenly misfired and I feared it might cut out completely. The reporting continued:

> 'His engine was heard to be misfiring and a sudden and most distressing attack of cold feet took possession of every onlooker, for even the French seemed quite keen on seeing the fastest machine win, recognising that the Sopwith is, indeed, the fastest thing on floats, and that a French win would be no sort of win to feel pride in.'

So far I'd been flying low, but I took the machine higher so that I would be able to turn into the wind and glide down should the engine fail altogether. For the next few laps my lapping speed was irregular, but as my engine gave no further trouble, I came down lower once again. *Lap after lap went by and the Tabloid kept going.* My race was made easier with the retirement of the two French competitors, Espanet in the seventeenth lap and Levasseur in the eighteenth, leaving only Burri and me in the air. Burri was flying much slower than I was, so I passed him several times, then he lost about thirty minutes for refuelling. I still had ample petrol to finish the race, and during the last six laps I reached an excellent regularity of speed.

> 'Finally amid cheers and great enthusiasm Pixton crossed the line for the 28th time.'

My race was over. I could hear shouts of *'Hurrah'*. I'd flown for two solid hours, but just before coming in to land I did a couple of extra laps to qualify for the 300-kilometre Seaplane World Record. The wind had risen, below me the choppy sea looked ominous and I had visions of turning over at the height of victory as I came in to land, but all went well. At the end of 30 laps, I made a good landing in the rough water. Mahl, who'd been out in a motorboat all the time I was racing ready for *eventualities*, steered towards me looking as pleased as punch and fixed a rope to the Tabloid to tow me back to the harbour. It was reported:

'No other Seaplane ever built has attained a speed of 90 miles an hour. No one, indeed, has ever exceeded 80 miles an hour hitherto. From this it is evident that the one apparently insuperable disadvantage with which every biplane was credited according to popular opinion, a lack of speed, has been removed once and for all.'

I left the cockpit and stood on a float while under tow to add extra weight to the front as the floats became half submerged and the tail dragged in the water when it was not moving under its own power.

I'd won easily!

I was very pleased with the Tabloid's performance. It had behaved beautifully in the air and on water, skimming across the surface as light as a feather and outclassing the rest of the field. Yes, I had won easily, even though some competitors had not yet started and were still at liberty to do so for the time was only 10.30am, but obviously no machine present would beat the Tabloid. It was also reported that Garros himself referred to *'the incontestable superiority of the little biplane, whose constructional qualities accord admirably with those of the pilot, who handled it with such a master hand...'*

Jacques Schneider welcomed me most heartily as I beached. *'Bravo, Monsieur Pixton.'* Sopwith warmly congratulated me, Harold Perrin and Harry Delacombe representing the Royal Aero Club shook my hand, a very, very nice man. A crowd gathered and a bouquet of flowers was placed on the nose of the machine, but as I glanced around I was astounded to see French designers dashing round the Tabloid with tape measures and rulers in their hands and, hardly able to believe my eyes, I asked Sopwith, *'Don't you mind?'* He did not. What appeared to the French as an unsuitable entry, even ridiculed, was no longer being ignored, and the French reporters were eager to get news of the flight first hand. Amongst them was the *Excelsior* correspondent who made quite a long report of the race and quoted me, though not always accurately:

> 'I have only been disturbed by the buoys showing the way because their blue colour mixed with that of the sea... It was the first race I have done (*of this kind*) and the first time I have gone round pylons... It was fairly easy, thanks to the balance of my plane. I took care not to waste time and shortened my turns as much as possible... At 100 kilometres (*more correctly about 150 km*) I had an anxious moment, the engine was not working well. But the excellent Gnome engine went well enough again for me to win, but I had a fright... I am happy that my first race (*first International Seaplane race*) was a win for England... But I am not a new pilot for I have flown the first plane (*one of the first planes*) built in England. It was a triplane whose beginnings were far from successful... We have improved since.'

A telegram came from the Gnome engine people, *'Enthousiates felicitations pour votre magnifique performance et pour votre belle victoire,'* And *The Aeroplane* correspondent present sent the news home:

> 'As I telegraph to you, Howard Pixton on the little Sopwith hydroplane with 100 horsepower monosoupape Gnome engine, British Integral tractor and doped with Cellon, defeated the pick of the world's fastest Water Planes in the most decisive manner. And right here I would like to say that the famous French and other pilots who are here, and have actually flown in the race, or at least watched from the shore, are among the first to give full credit to the constructors and pilot of the Victorious Machine, and from no one does one hear higher or more unstinted praise than from these good sportsmen.'

I sent a postcard home to tell my mother that I had won the race and that I would be returning Wednesday and home Friday. My top speed, though handicapped by the drag of the floats, worked out at 93 miles per hour and 86.75 was declared officially as my average speed, an average which could have been higher but for the trouble which arose in the fifteenth lap. Mahl discovered the trouble was caused by a broken cam on one cylinder which had put it out of function so that I'd been flying on eight cylinders during the last half of the race, but even so the Tabloid was now looked upon as *the fastest waterplane in the world*. I had set a new World Record for Seaplanes over 300 kilometres, but it was more than that.

This was the first international coup by any British aeroplane.

I'd developed a violent thirst flying non-stop for two hours, and when Schneider invited me to celebrate my win in the Sporting Club close by, I was pleased to accept. Though generally not a drinker, when asked what I should like to drink, I replied, *'Thanks, mine's a half a Bass,'* I was teased unmercifully. 'After a glorious win, and luxurious hospitality in a country of wines and champagne, all you want is a beer...!' As I peacefully drank my Bass, I found myself surrounded by people offering their congratulations. 'A remarkably fine performance, *I've never seen anything quite like it*, a marked triumph.' It was nearing noon when Burri came in. He was the only competitor other than myself to finish the course, and his time was three and a half hours, average speed 51 miles per hour, a considerably slower flight than mine. Neither of the Americans, Weymann or Thaw had attempted a start. Levasseur resumed his flight hoping to finish third place, but did not make it. True to his word, Garros did not even attempt to beat my record flight, but in the afternoon he had given his place to Prévost who wanted to fight for the honour of France, but after a testing lap he too had to retire on breaking his propeller in the bad sea as he made the compulsory landings. Although the French had thought Garros would wrap rings round the Tabloid, we had defeated them and as much as the French newspapers tried to say it was not a French defeat since both winning aircraft were equipped with 100-horsepower Gnome engines, the fact remained, we had won.

France was forced to admit defeat.

\multicolumn{4}{c}{**20th April 1914**}
	Pilot	Nat	Plane
1st	PIXTON	GB	Sopwith Tabloid floatplane
2nd	BURRI	SW	FBA Flying Boat

20th April 1914

THE SCHNEIDER TROPHY SEAPLANE RACE

	Pilot	Nat	Plane
1st	PIXTON	GB	Sopwith Tabloid floatplane
2nd	BURRI	SW	FBA Flying Boat

That same evening, by invitation of Jacques Schneider, we attended a friendly banquet held in the Empire Room of the Hotel de Paris, and among the many guests were Lieutenant Burée, ADC to the Prince of Monaco, Lieutenant Destrem of the French Navy, Monsieur Boris of the French Naval

Engineering Service and Monsieur Georges Prade, Clerk of the Course. And what a grand feast we had, fit for a queen. Schneider was an exceptionally pleasant man, delighted that his race had been such a success this year, even though we had beaten his fellow countrymen and were taking the trophy out of France. Needless to say, during the course of this eventful day a group of pilots, including myself, went to the Casino but almost immediately I lost a few fivers at the gambling tables, so I wasted no time in leaving the famous Casino while the going was still good.

Next day, His Serene Highness Prince Albert I of Monaco formally presented us with the trophy. It was magnificent, made of silver bronze and designed to show a winged female spirit, Zephyr, god of the gentle west wind, sweeping from the skies to kiss the faces of figures blended in a crest of a wave, the base of which was ornate with octopuses and crabs. It was some size too and would look grand in any place it sat. Britain, as the winning country, now became responsible for the trophy until the next Schneider Race which would be organised by us. What I had done by winning the Schneider Trophy was to raise Britain to the top position in aviation. Sopwith had instant recognition as being the foremost constructor in the world and France now took second place.

The centre of aviation shifted to Britain because a British plane had won an international race.

The Wright brothers' achievement in getting an aeroplane off the ground went almost unnoticed by the British Press. So it was with the Schneider win. Although it was just another race, it was an important race and the Press were slow to realise the full significance of the victory. Publicity came mainly from Flight and Motor Journals that backed their stories with ample use of photographs. '*The Return of the Conquerors... The First British Victory... Ahead of the Rest of the World... A historic moment in British aviation...*' And *The Aeroplane*, wrote.

> 'At last, and for the first time in history, a British built aeroplane has won a classic race, the Schneider Trophy having been captured by Mr Pixton on a Sopwith Water Plane on April 20th.'

There were two great races established, the already famous Gordon Bennett Trophy of America and now the Jacques Schneider Trophy of France, its marine equivalent. Grahame-White was the first to win the Gordon Bennett Trophy for Britain, but it did not rate as high as the Schneider Win since it was won on a French machine, the Blériot. Our next step now was to develop British aero engines, to develop one as good or better than the popular Gnome, and already the Government had made a start by organising an Aero Engine Competition, which was at present underway. Further reports appeared giving readers a clear picture of the significance of this win.

> *Flight*, 'It is almost impossible to overrate the importance of the winning of the Schneider Trophy by the Sopwith machine piloted by Mr C.H. Pixton. The French have made no secret of the fact that they consider the British Victory as a most serious blow at the prestige of French aviation.'
>
> *Daily Telegraph*, 'Mr Pixton's success, in addition to raising the prestige of the British Industry as perhaps nothing else could have done, is significant of many things.'
>
> *The Motor*, 'A Big British Victory. The performance of the British Sopwith Waterplane at Monaco in winning the Jacques Schneider Trophy is one of particular merit. If Great Britain takes a little time, compared with other countries, in getting into her stride, she scores in the long run.'
>
> *The Car*, 'At last a British-designed, British-built, and British-piloted machine has made good in a big International Event, and for the first time in the history of aviation we have beaten the Frenchmen thoroughly and most convincingly.'
>
> *Aeroplane*, 'Mr C. Howard Pixton, known to many of us owing to his early flights on the little 35-horsepower Avro, and his excellent handling of Bristols, including the winning of the Manville Prize, as a first class flyer, but now for the first time attaining worldwide fame as the winner of the big International Event.'
>
> *The Surrey Herald*, 'Mr Howard Pixton, of 3 King's Road, Walton, had the satisfaction of winning the Jacques Schneider Trophy for water planes at Monte Carlo. Mr Pixton was on a Sopwith biplane with Gnome engine, and he won The Trophy and £1,000 in addition, at an average speed of 86 miles an hour. With the exception of M. Burri (Switzerland) who was adjudged second, all the others had to be towed home.'

C.G. Grey had his say in expressing his disappointment of the slowness of the Press for their lack of enthusiasm.

> 'The real interest of the British press in British aviation is shown by the fact that Reuter's brief telegram seems to have been sufficient for

their enthusiasm over our first International Victory. *The Mail*, which is so fond of impressing on its readers all it has done for aviation and its faith in British water planes as our safeguard in time of need, omits to draw any lesson from the Sopwith's performance. It gives just three times as much space at the head of a column to a blind man's five minute trip on a foreign aeroplane, while the Schneider Trophy win is stowed away at the bottom of a page.'

Flight took it up too:

'An Unnoticed British Triumph. It is passing strange that so little attention has been given by the Press at large to the winning of the Jacques Schneider Trophy at Monte Carlo by a British Seaplane, the Sopwith piloted by Mr Howard Pixton which was the only one to complete the overseas course of 150 nautical miles without trouble. The Press has done so much for the cause of British aviation that it might reasonably have been expected that something more than an average of half a dozen lines would have been devoted to so notable an achievement. We are not finding fault, for to do so would be to manifest ingratitude for all the Press has done for the development of the science, but it is impossible to allow the matter to pass without remark. Ever since aviation came into being we have been accustomed to regard ourselves as lagging behind others in development, at any rate until quite recently, with the exception of the win of the Cody machine in the Military Trials of 1912, the British-built machine has almost invariably in the big events had to be content with a record of "also ran", and the Press has not been slow to point it out. Now that we have really made substantial progress, so substantial that the tables have actually been turned, the fact apparently escapes all notice save that contained in the bare reports.'

Well, so much for the strong feelings! I'd simply won the 1914 Schneider Race but France, who had been so well-ahead of Britain having trained 344 pilots by 1910 to our 50, and who had claimed so many of the *Daily Mail* prizes, and had done so well in the British Military Aeroplane Trials of 1912 was now, apparently, left quite stunned.

Sopwith took advantage of the win and issued a full page advertisement on the Sopwith Company with the heading 'The Schneider Trophy Race was won on a Sopwith Biplane, Piloted by Mr C. Howard Pixton.' Two photographs of the machine during the race illustrated the advertisement, and further details of the company were included. Another full page advertisement by the Gnome Engine Company appeared. 'The Schneider Trophy Race result, 1st Pixton on Sopwith plane, 100hp Monosoupape Gnome Engine, beating all speed records for Hydro-aeroplanes up to 300 km. 2nd Burri on FBA Flying Boat, 100hp Monosoupape Gnome Engine.'

So impressed was Major Wood with the Gnome engine that he ordered a large quantity for the Vickers Gun Bus, and the Aero Club received a warm letter from the Aero Club of France, typical of the friendliness of the French people, 'We have great pleasure in sending you our most hearty congratulations on your recent brilliant victory in the Jacques Schneider International Maritime Race. We had hoped indeed to retain this valuable Trophy, but it is a consolation to us that it has been won by a nation who is the friend of France, whose Aero Club entertains the best relations with you.'

Jacques Schneider presented a plaque to the Aero Club to commemorate the British victory.

CHAPTER TWELVE

With Sopwith after Winning the Schneider Trophy

I couldn't believe my ears when I stepped into Brooklands a few days after the Schneider win and was ironically accused of foolhardiness. The Brooklands Manager, Frank Wright, approached me. 'It is considered by the Brooklands Flying Committee that your particular style of flying is such that it constitutes a danger to the public.' I told him that for the main part I'm testing War Office machines for the Sopwith Company, and cannot imagine how my flying is thought to be dangerous. He replied that they were concerned about my exhibition flights during the Race Day Meetings, 'I fly to entertain the public,' I said, 'who appear to like my flying, and it is in no way dangerous.' He told me seriously that my flying was alarming. I said 'I shall take greater care,' but the next day I received a letter dated 27 April 1914:

> 'Dear Sir,
> Flying at Brooklands to the danger of the public. With reference to my conversation with you yesterday afternoon, I beg to confirm what I said to you regarding the dangerous nature of your flying when you promised to exercise greater care in the future. As you are doubtless aware, the Home Office have appointed or contemplate appointing, Special Inspectors to visit Flying Grounds and look out for dangerous flying, and if the Management were to take no steps to put a stop to it, the Home Office would undoubtedly exercise their powers and close the Ground to the public. An acknowledgement of this letter will oblige.
> Yours faithfully, Frank Wright.'

Not long after we returned home the Schneider machine, minus its floats and refitted with wheels, appeared at Brooklands and caused a considerable stir. Then came the luncheon given by the Royal Aero Club, at the Royal Automobile Club, Pall Mall. The Marquis of Tullibardine, Chairman of the Club, presided. It was a pleasant occasion. Among the guests were Colonel

Holden, Vice Chairman of the Royal Automobile Club, Brigadier General Henderson, Head of the Air Department at the War Office, S.F. Edge, President of the Society of Motor Manufacturers and Traders, Stanley Spooner of *Flight*, and C.G. Grey of *The Aeroplane*. Others present, many of our friends, included Harold Perrin, Harry and Roy Delacombe, Major Lindsay Lloyd, Henry Morris, C.C. Turner, Holt Thomas, Cedric Lee, Harry Preston, Pemberton Billing, Lieutenant Spencer Grey, J.H. Spottiswoode, Bell, Hucks, Valentine, Gordon England and many more.

Harry Delacombe, who'd been with us at Monte Carlo, played a large part in organising the reception. We first met at the 1910 Blackpool Meeting, and at one time he was with Bristol in charge of the pilots... a friend of pilots, always interested in flying but never flying himself except as a passenger. He was also a great pianist. I'd often heard him and his wife play at Sopwith's home, and they played beautifully. One thing he was most ardent about was London. He loved London. I've never known a man with such passion for the city and he could not understand why anyone should want to live outside it.

I was delighted to find myself sitting by S.F. Edge, the 'king of speed' motor ace of Brooklands, a representative of the makers of the great car, the Napier. I mentioned to him that I was an avid admirer of his and made a point of watching him whenever he was racing at Brooklands, a case of hero worship. He thought that this was amusing coming from me but it was mutual admiration, I enjoyed his racing, he enjoyed my flying and remembered the days when I was always dropping into the sewage beds. We recalled the Brooklands to Brighton Race when I went off course to Plumpton on the Roe biplane. 'You passed over my home in Sussex' he said 'and I thought you were going to alight in one of my fields.' I had been flying so low but landed on Plumpton Race Course just about a mile away.

After the meal the Marquis of Tullibardine rose and proposed the toast, 'Mr Sopwith and Mr Pixton.' He said:

> 'We are here to congratulate an English constructor and an English pilot on Britain's first international aviation victory on a British aeroplane. I refer, of course, to the splendid performance of Mr Sopwith and Mr Pixton in winning the Schneider International Maritime Race at Monaco last month. We must not forget that Mr Grahame-White won England's first International Race, the Gordon Bennett Trophy, in America in 1910 on a French machine. True, in the Sopwith machine the motor was French, but we need not grudge this share of victory to our good friends across the Channel and it is only fair to recollect how much the Gnome motor has assisted all constructors and pilots in

advancing the science and practice of aviation for the past four years. We can only hope that the Government Air Engine Competition now in progress may produce one or more British motors the equal, if not the superior, of anything designed and manufactured abroad.

Before we heard of Mr Sopwith as a constructor, he was world famous as a pilot, first achieving fame by his victory in the Baron de Forest prize of £4,000 on an all-British Howard Wright biplane. Soon after he established a reputation in America as one of the world's greatest race flyers, and gave us a taste of his ability in this direction by his many successes in Handicap Races at Brooklands, followed by his victory in the first *Daily Mail* Aerial Derby round London. For the past eighteen months, however, he has deserted the pilot's seat to devote his whole time to the manufacture of aeroplanes, and combining his practical experience with his engineering knowledge, ranks as one of the foremost Aeroplane builders in the world.'

He had a few more things to say about Sopwith, then spoke of my flying. 'Mr Pixton took up flying in the early days of aviation, and since that time he has justly earned the reputation of being one of England's most capable pilots.' I had given Britain Air Supremacy. Sopwith replied in a humorous vein, speaking of the trials in the Southampton Water and emphasising the wonderful qualities of the Gnome engine, adding that there is a saying, 'A tea tray would fly with a Gnome in it'. He continued:

'The actual machine which secured the Trophy is a direct outcome of building machines for the Government to General Henderson's requirements who has done so much for aviation. A high tribute must be made to the splendid skill of Mr Pixton who flew in a masterly fashion, circling the pylons with barely a foot to spare in a manner that drew the unbounded admiration of even the most famous French pilots in Monaco. *I have never seen a race flown as Mr Pixton flew for the Schneider Trophy.*'

I replied briefly, giving due regard to Sopwith as a designer and to the excellence of the design of the Tabloid. 'Any pilot could have won on the machine I flew.'

Colonel Holden proposed a toast to the guests, especially mentioning General Henderson and S.F. Edge. General Henderson said a few words:

'How pleased I am to be here to congratulate my friend, Mr Sopwith, and my own former instructor in flying, Mr Pixton, on their joint victory. Mr Sopwith and other British makers have produced aeroplanes better than any in the world. And foreign pilots never were better than British pilots. They merely had more opportunities. In both respects, we are now distinctly ahead of the rest of the world and I hope the next International Victory will be that of the British engine.'

S.F. Edge replied. 'I am very gratified to know that we have succeeded in the interests of Britain in building a machine which has such flexibility as the Sopwith, greater flexibility in speed than any other aeroplane in the world.' The health of the Chairman was proposed, followed by a vote of thanks.

Gordon England and I chatted about things in general, I confided to him, 'You know, Eric, I think I will give up flying and go into something new, a business of my own perhaps.' 'But why? You're at the height of success,' he stated. 'Flying's for the young. Besides it's getting too commercial,' I replied, at which he was rather surprised.

I'd started the tests on the Tabloids ordered for the War Office as they came in from the works, flying them at Brooklands then taking them over to Farnborough. While delivering the third or fourth, however, I thoroughly disgraced myself at Farnborough. Being nothing more than too clever, I landed slightly down wind and came to rest near the Aeronautical Inspection Department's sheds, but instead of the machine coming to a complete halt, the tail continued to go forward and the whole machine somersaulted over, so that I found myself for the second time strapped in an upside-down machine. Little damage was done, a broken skid and a buckled wheel, but I should have known better...

The larger *Bat Boat* after the style of the *Bat Boat No 1* was to go to Germany and I did the tests from the sheds at Woolston, Southampton, where it had been assembled on Pemberton Billing's premises. I flew all the way to Netley, a distance of nearly 2 miles, without touching the controls as it flew absolutely straight and was remarkably smooth, another good aeroplane by Sopwith. Soon afterwards a German officer came to Britain to collect it, and I was to fly with him around Southampton Water to ensure he was completely satisfied, but while talking to the commanding officer at the Calshot Naval Air Station, he suggested that I should not bring the German guest too near the station. Flying, anyway, was restricted over the area, but I asked, 'Why ever not? There's nothing to see.' 'That's why,' was the reply.

Between testing machines, I had been teaching Mahl to fly. He had done a lot of work for us at Monaco and I was immensely pleased when he passed his tests during May, securing Licence No 784. His desire to become a qualified pilot had at last materialised, and so soon after the Schneider Victory. Now Sopwith had three pilots, Hawker, Mahl and myself.

Hendon had grown into a very successful showground and was constantly well-spoken of by journalists, 'A resort of fashion... a veritable Ascot in London... a splendid health-giving interesting, pleasurable worry-forgetting rendezvous, with everything that can possibly be thought of for the comfort of visitors... And the World of Fashion in London has received it with open arms... Splendid and comfortable tea pavilions, little striped red and white garden tents scattered about, each with its wooden floor, and its dainty tea service prettily and invitingly set out... Plenty of walking space, one can promenade if so minded, thousands of comfortable chairs, arm and otherwise, for the weary... There is plenty of fine flying, music, fashion, sport, interest, comfort and fresh air...'

Although under unfortunate circumstances, we have a lot for which to thank the Frenchman, Marcel Desoutter, who did his flying at Hendon. During 1913, young Desoutter had had a serious accident on a Blériot when it crashed. The front of the machine crumbled and squashed one of his legs below the knee and surgeons at King's College Hospital, anxious about the condition of the leg as tetanus had set in, were forced to amputate it. Desoutter himself was so concerned that a smash, such as his, could result in amputation that he gave a warning to all pilots. 'If you ever have a smash and soil gets into your wounds, have an injection for anti-tetanus serum at once. It is expensive and can only be obtained from France, but it'll be worth it.' To the relief of many pilots, Burroughs Wellcome began manufacturing the serum, no doubt having noted Desoutter's comments on its unobtainability in Britain. After the amputation, many thought Desoutter would not fly again, but no. As soon as he was fit he was in the air, the first pilot to fly with one sound leg only, and was very much admired for his pluck and courage. Because of his determination to combat his handicap, he refused to be content with the conventional artificial limb made of wood, which was ungainly and heavy, and with the help of his brother designed a better limb, weighing only two pounds, of aluminium covered with leather. It was a great improvement and easier to wear. The Desoutter brothers patented the idea and started a business in limbs.

Our next main flying event was the 1914 Aerial Derby, an annual race around the outskirts of London organised by Hendon and sponsored by the *Daily Mail*. I was to fly in it for the first time. The Derby, now in its third year, was one flying event of the year of all round interest. It started and finished at Hendon, a circuit of 94^1/$_2$ miles with turning points at Kempton Park,

Epsom, West Thurrock, Epping and Hertford. This year promised a great display of racing with twenty machines entered. I was quite looking forward to the event, but things did not go according to plan. The race was being held on Saturday, 23 May. Excellent weather promised a good race, then there was a sudden change. A thunderstorm raged the night before and on the day of the race clouds hung low as I flew in from Brooklands. I practically lost my way because of the haze and could not see my regular landmarks. It was only when I sighted the waters of the Welsh Harp that I knew I was on course to Hendon. I arrived at 11am and people were already assembling although the race was not due to start until 11.45am. While waiting, we gave exhibitions of flying for the benefit of spectators, but no one was happy about the weather. Only nine of the twenty contestants were present, and it looked doubtful whether the rest would appear.

We were racing for the *Daily Mail* Gold Cup and the Shell Trophy, and cash prizes of £100, £75, £25 in addition to £200 for the fastest time. It was a handicap and I was to start from scratch on the Tabloid Scout flying against the Avro Scout, Bristol Scout and the new Vickers Scout. In other words, the Tabloid Scout was considered to be the best machine entered. By 2pm it was raining steadily and visibility had not improved, and at 3pm Queen Alexandra arrived with the Empress Marie Feodorovna of Russia and Prince Aage of Denmark. The crowd had swollen to about 30,000 but the weather showed no signs of clearing and when Lord Carbery and Goodden put on a show of looping it was only just perceivable through the dense sky. I was doing some low flying for the public and came in to prepare for the start. As the machine's landing speed was very fast, about 30 miles per hour, I damaged the chassis on the bumpy surface of Hendon's ground. I was a bit annoyed, but that was that. The machine had a good shake-up and the damage could not be repaired in time for the race, but shortly afterwards an announcement was made, 'Due to bad weather, the Aerial Derby is postponed to 6th June.'

There'd been no sign of Hamel, a fellow competitor for the Derby who was expected from France having left the day before to collect a new 160-horsepower Morane-Saulnier especially for the race. He had reached Villacoublay all right and left on the new Morane, then was reported seen at Hardelot on the French coast. Hamel was wearing an inflated bicycle tube around his body, not an entirely unknown habit of pilots crossing over water, as he left for the English Channel, but nothing more was known. Hamel was certainly very familiar with the Channel having crossed it many times and had only recently announced his intention to cross the Atlantic, but the days went by, naval ships searched the Channel and finding no trace of him nor his machine, he was presumed lost. A few weeks later his body was picked up by a French fishing smack and was identified by the clothing, articles found in the pockets and the inflated tube, but the body was not brought ashore for burial because of its decomposed state. Hamel, the son of a Danish surgeon, was born in London and educated at Westminster School, then took up flying having received a French licence and also an English one in 1911. I'd flown against him on several occasions, a most popular young man. We all knew him very well, another very sad loss.

Saturday 6 June. I got ready once again for the 1914 Aerial Derby, but the weather was just as foggy as it had been a fortnight ago. I arrived at Hendon after noon flying a 80-horsepower Sopwith Scout since Barnwell, who was to fly for Vickers, switched to the Schneider 100hp machine at the last minute and was competing against me. Mahl came along on a three-seater to cheer us on, but many of the competitors did not even make it to Hendon. It was touch and go whether there would be another postponement as the bad weather persisted and showed no signs of clearing. We lined up. This time eleven machines out of twenty-one were present. Barnwell was now the scratch man, and because of the terrible conditions, we were reported as being the 'Eleven Brave Men.' A few who were taking passengers were given extra time for the weight they were carrying in addition to the time advantage given to their machines in the scaled handicap. First man away at 4.16pm was Bjorkland followed by Birchenough and passenger who had a forty-eight minute benefit, a fantastically high advantage over such a short course which should take about one and a half hours if flown at 60 miles per hour, then Carr and passenger, Verrier and passenger, Strange, Alcock, Brock, Noel, Carbery, all with time advantages, and myself with a six-minute advantage over Barnwell. One after the other we went out at one-minute intervals, each of us making a circuit of the aerodrome before getting away.

Our Sopwiths caused a great deal of excitement, they were fast and the crowd knew it. Despite the thick mist, we all got away and were flying uneventfully. Soon I'd passed six machines and was behind the leading one, Brock, and then I even passed Brock but decided to quit. It was ridiculous to continue racing under such conditions. I came down, but I wouldn't have got very far anyway, for I found the petrol pipe to the carburettor was just hanging on and was spurting petrol. Alcock also became tired of the conditions and flew back to Brooklands. Barnwell got completely lost. Brock was declared the winner, Carr second, Verrier third, the only three to get around completely. Barnwell or I, should have won as we had the fastest machines and we'd both passed Brock early in the race. However, it was exceedingly clever of anyone to have got around at all in such appalling conditions. Brock, the winner, was an American who'd come to England in 1912 to join the Deperdussin School at Hendon. On getting his licence, he became the school's instructor until it closed down. He'd just revisited America and was not very enthusiastic about flying there because of the Wright Company's monopoly. 'The public has lost all interest in aviation in the States. There's no scope there... *That company is the bugbear of aviation!*' His opinion of Britain was, 'She's a cute little toy country... .' So ended the third Aerial Derby, Hendon's popular race with a history of winners building up, Sopwith in 1912, Hamel in 1913, and now Brock in 1914.

6th June 1914	
Aerial Derby **Eleven Brave Men**	
Bjorkland	
Birchenough	
Carr	2nd
Verrier	3rd
Strange	
Alcock	
Brock	1st
Noel	
Carbery	
Pixton	
Barnwell	

The day after the Derby, Harry Hawker arrived in Southampton by boat from his six-month holiday in Australia, and his Tabloid was expected back at Brooklands sometime in July. His feet had barely touched the English soil when he was up in the Schneider plane with its re-attached wheels. Then he was racing again, first competing in the London – Manchester – London Handicap on 20 June. I was not competing, but most of the active pilots of the Aerial Derby entered, Birchenough, Carr, Strange, Alcock, Brock, Noel, Carbery and Hawker, the favourite and the scratch man. There were eight starters out of fourteen entrants since again the weather was foggy. Hawker only reached Coventry and came back feeling unwell but Brock, who seemed his best in foggy conditions, went on to win, flying the 324 miles in four hours and forty-two minutes, an average of 69 miles per hour. Carr was second,

Alcock third, again only three completed the course. The next race Hawker entered was the June Brooklands Handicap. In fact, we all entered, Hawker, Mahl and myself, but the handicapping was once again so poor that Hawker and I on the Tabloid Scouts did not stand a chance of winning, and Mahl came in first on a standard Sopwith.

Hawker had done quite a bit of looping on the Tabloid and arrangements were made for him to give regular looping displays at Brooklands each Sunday afternoon. He looped with ease, but on the evening of the June Handicap he got the 100hp ex-Schneider Tabloid into a spin. We watched him as he dived to commence a loop but on reaching the top of the curve the machine fell sideways, turned nose down and spun vertically as it lost height from about 1,000ft, it flattened out at 200ft from the ground but dived again and disappeared into St George's Hill Wood. We were horrified, *It must be the end for Hawker.* I took the nearest available machine to locate the wreckage, but could see nothing because of the denseness of the wood, and returned fearing the worst. Downhearted we hurried off on foot to search for the wreck, but as we were going along the road taking us towards the spot where we believed he'd crashed, a motorcycle drew alongside us and there, riding pillion was Hawker himself, grinning broadly. Weren't we pleased to see him! *A spin meant death!* 'I hit the trees, hung suspended for a few seconds, then sunk through the boughs to the ground,' he said.

He wanted to know how to recover from a spin and thought he knew what to do. He would now use the 80-hp Tabloid and put it into a deliberate spin! Next day Hawker risked his life to find out exactly how one could recover from the dreaded spin which was still known as *Parke's Dive*. Although we were looping and doing other skilled things in the air, no one knew quite how to stop the spin once it had started. Parke, who'd had a lucky escape, did not completely understand what happened in his recovery, neither did aviation experts who'd tried to explain its technicalities. The Public Safety & Accidents Committee had even taken it up with the National Physical Laboratory to investigate the cause and remedy, but no satisfactory explanation had been forthcoming. I'd never been in a spin but it was a weird thing. Once it started there was no means of regaining control, the wing would go down and the more you tried to correct it, the worse you made it so that the spin always ended up in a crash which usually meant death. To date only three pilots had survived the spin, Raynham on an Avro in 1911, Parke on the enclosed Avro in 1912, Pickles on a Champel Pusher in 1913. Pickles crashed at Hendon during the second Aerial Derby with Mrs Stokes as his passenger and escaped with a broken leg and minor injuries, but Mrs Stokes was unconscious for three days. Parke recovered from his spin by chance at the Military Trials and I'd seen Raynham's spin at Brooklands before that. They were all very lucky to be alive.

Physically Hawker was unimpressive, not a man always in the best of health but a wonderful pilot, one of the finest I'd ever met and one of the bravest. No one could dissuade him from taking up the Sopwith Tabloid the next day. No matter how hard we tried to influence him, it made no difference. '*It's plain suicide, Harry.*' And someone said, 'The Scout will never come under control again because of its short tail,' *but Hawker was ready*! I felt sick with anxiety wondering whether this would be his last flight, but he looked as though he was off on a joyride as he cheerfully left the ground.

He rose higher and higher until he was at about 1,500ft, then it started, a deliberate spin, *the first deliberate spin in history.* Hanging on its tail, the machine slowly rotated and down he went, but nothing was happening, then quite suddenly the turning stopped and the spin became a straight dive. *He'd done it..*! He made a perfect landing and we went to meet him, we could hardly credit it, '*What did you do?*' He replied, 'I did nothing, absolutely nothing. I took my hands and feet off everything and, as I thought, came out of the spin and found myself in the dive.' Gradually we understood what had happened. All controls must be centralised then wait, the spin stops and all that is left to do is the recovery from the ensuing dive, a normal flying manoeuvre. It was quite the opposite to what we imagined, there was no need to pull the throttle back to try to get the nose up or to try to turn out of it. He did all the obvious things, nothing happened *until he neutralised the controls!*

Hawker's courageous deed, witnessed by many and known among all of us, was never published. Not a word of it appeared in any of the newspapers or flying magazines, but without a doubt Hawker was the first person to discover the remedy of the spin, and all credit must go to him, and him alone, and certainly not to Parke who had no idea how he managed to get out of his spin!

Racing was becoming more sophisticated. Bigger races were being arranged, and more frequently. The first London - Paris - London Handicap was held on 11 July only three weeks after the first London - Manchester - London Handicap. We did not enter but out of the fourteen entries there were seven starters, Hearn, Renaux, Noel, Brock, Carr, Carbery and Garros. Visibility was poor once more and only three finished. Again Brock won, he flew the course of 508 miles in seven hours, an average speed of 71 miles per hour, and became the first man to travel by any means of transport from London to Paris and back in one day. This remarkable American flyer had now won three major British races one after the other in just over a month. The Gnome Engine Company took advantage of his wins as his Morane was fitted with one of their 80-horsepower Gnomes, and were now advertising his successes alongside my Schneider win as, 1st in The Schneider, 1st in The Derby, 1st in The London - Manchester - London, 1st in The London - Paris - London.

We were preparing for the *Daily Mail* £5,000 Round Britain Seaplane Race postponed from last year and due to commence on Monday, 10 August 1914. The rules were the same except that it was not to be a passenger-carrying event. I was entered and would be flying the new Sopwith Bat Boat, which was rated as one of the finest sea machines of the day. The engine was a powerful 225-horsepower Sunbeam and the tanks held 70 gallons, enough for five hours duration. I had a very good chance to win here. Mahl was entered too, it was to be his first big race. The 1914 Gordon Bennett International Race was also due to take place in France during September. Roe, Bristol, Vickers, Cedric Lee and Sopwiths all entered to represent Britain. Britain was also thinking about The Third Schneider International Race due to take place early in 1915, and it seemed that someone would soon take the *Daily Mail's* £10,000 Atlantic Crossing prize. Hamel was one of the first to say he would attempt the crossing, his large transatlantic steel machine constructed by Martin & Handasyde at Brooklands had become a familiar sight at the ground, but since his death Lieutenant Porte had announced his intention to make the crossing during August or September on a Curtiss Flying Boat, which was ready to fly. He planned to leave from Newfoundland. Then the Round the World Race, balloon, or aeroplane, was arranged for May 1915 with prizes of $100,000 (£20,000), $30,000, $20,000, the greatest flying prizes ever offered!

Then War stopped everything.

All Air Races were postponed as the 4th August 1914 brought this unique period in the history of aviation to an abrupt end. The first signs we may be involved in a war was made apparent by a notice issued by the Marquess of Tullibardine, Chairman of the Royal Aero Club. 'Owing to the grave state of affairs on the Continent, it is possible that the British Empire may be involved in a European War. In such an event, the assistance of every able-bodied man might be required, and it is felt that no class of the community could be more of use to the Naval and Military Authorities than the flying man.' And a few days later we were at war. I was in Southampton at the time. The first indication I had that war had been declared, was when I came face to face with a newspaper poster of the Kaiser under which was written...'Stark, Staring Mad.'

Crowds gathered at recruitment offices set up all over the country, pilots enlisted and constructors got ready to meet the demands of war, the drilling of thousands took place on Salisbury Plain and other open areas. Brooklands was taken over in its entirety by the Royal Flying Corps who turned it into a military air base. Those who did not go into service continued to work there and the Blue Bird Café was converted into a canteen with Mrs Billing left in charge, but she didn't stay for long. Hendon was made into a Naval Air Base and the schools there, though operating in the usual way, were put under official surveillance. Porte, fresh back from America cancelling his postponed Atlantic crossing, was placed in command and Grahame-White became his assistant but, unlike Brooklands, Hendon was to remain open to the public during the seasons until further notice. Captain Wood of Vickers was called back to his old regiment, Jullerot returned to France with several other French pilots who'd made Britain their home, and Coanda thought he would return to Roumania as it seemed likely that his country might be brought into the war. Merriam, Bristol's gifted instructor, was appointed to the Royal Naval Air Service to instruct pupils at Hendon, Harry Delacombe of the Aero Club became responsible for recruiting mechanics there, and Major Lindsay Lloyd, Clerk of the Course at Brooklands, joined the Army Transport Section as did many pilots who'd given up flying, to be of service with their cars at home and at the Front.

Britain had trained a total of 862 pilots since the beginning of flying, some were dead, some were pilots from overseas trained in Britain, but as changes were made many a happy pilot mourned the loss of his pet machine and turned to face war.

The friendly days were over.

It was rather an unfortunate coincidence that the last peace issue of *The Aeroplane* had pictures of German pilots holding the world height and duration records for its front page! It was captioned, *'Our Betters'*.

CHAPTER THIRTEEN

War 1914–1918

His Majesty's Government informed the German Government on August 4th, 1914, that unless a satisfactory reply to the request of His Majesty's Government for an Assurance that Germany would respect the neutrality of Belgium was received by midnight of that day, His Majesty's Government would feel bound to take all steps in their power to uphold that neutrality and the observance of a Treaty to which Germany was as much a party as Great Britain.'

That's how it started as Germany prepared to advance into France via Belgium and Luxembourg. The Press kept us informed. 'The German plan is clear. It is also unalterable, now that matters are so far advanced. Everything is being staked upon the success of a movement, in enormous strength, through neutral territories of Luxembourg and Belgium.' Much credit was paid to the aeroplane.

'The French and her allies know where the main strength of the enemy is massed and what its intention is, they even know approximately what numbers and what sort of troops are being brought up against them on the ascertained line of advance. No defence could ask for more. That they enjoy this knowledge is undoubtedly due in the main to the flying men.'

British pilots were sent to bases in France and a French correspondent writing from Paris expressed, 'Everybody is enthusiastic here and impatient to go, and we are absolutely sure of Victory with the help of the English.'

The aeroplane was being used in war! To begin with they were used for reconnaissance only, air battles came later, and the Tabloids I'd flown over to Farnborough during the summer were among the first to be seen at the Front. Most people thought it would all be over by Christmas, but C.G. Grey predicted, 'It appears that this country is inevitably committed to take part in the greatest war the world has ever seen.'

At home, Mr McKenna, Secretary of State, issued an order prohibiting Cross Country Flying:

'I prohibit the Navigation of Aircraft of every class and description over the whole area of the United Kingdom and over the whole of the coast line thereof and territorial waters adjacent thereto. This order does not apply to Naval or Military aircraft or to aircraft flying under Naval or Military orders, nor shall it apply to any aircraft flying within three miles of a recognised aerodrome.'

The public were warned to keep clear of Air Stations by notices posted outside them such as:

'All persons are hereby warned that all Army Sentries at the Waterplane Station or on the foreshore have orders to challenge once, and if not instantly obeyed to fire. On the order to halt, they must immediately do so. They will approach the Air Station at night at their own peril.'

A few days later this appeared in *The Aeroplane*:

'Charles Carroll, who at one time was an assistant to the late Mr S.F. Cody, was shot by a London Territorial Sentry on August 20th. He was examining an Aldershot railway bridge, and was challenged six times by the sentry before the latter fired. Carroll, who was almost stone deaf, died next day.'

The sad news of the first British airman to die in action reached us during the first weeks of the war, Vincent Waterfall. Vincent was with his observer, Lieutenant Bayly, during a reconnaissance flight on a Avro over Belgium. We knew him very well, they both died in Enghien, cause of death, 'apparently shot down by Germans'. Vincent Waterfall had been very popular at Brooklands where he flew for Martinsyde, always cheerful, always teasing Mrs Billing of the Blue Bird, and the news of his death cast a gloom over all of us who had known him at Brooklands.

I continued to work with Sopwith, testing machines for the Royal Flying Corps, but the days passed slowly as we waited for war production to get underway and I was not fully occupied, and for this reason decided to leave Sopwith. I left at the end of September 1914 to join the AID. The AID, short for Aeronautical Inspection Department, was a new civilian testing and inspection establishment at Farnborough, set up in January 1914 after a series of fatal accidents, to supervise all aeronautical construction. Manufacturers wanting to supply the RFC with machines submitted details and, if approved, a prototype was put in the hands of the AID to undergo tests then orders were placed. AID Inspectors kept an eye on production and the final testing was done at Farnborough. Before the AID came into existence, manufacturers submitted their machines to the Royal Aircraft Factory at Farnborough, but it was practically a direct competitor since they produced experimental machines for the Government based mainly on proven designs.

This is how the following breeds came into being, **FE** Farman Experimental, **BE** Blériot Experimental, **SE** Scout Experimental, **RE** Reconnaissance Experimental. I was engaged as an assistant inspector and test pilot in the Flight Delivery Section where my duties were to test and report on all air-delivered machines. Sometimes I went to the aircraft works to collect the machines, sometimes they were flown in by the company themselves. The work was not unpleasant and I was kept busy. Fulton was my boss, a very pleasant man, gentle in manner and determined in his ways, who held an early Flying Licence, No 27. He'd taken his tests just a month before I took mine in November 1910. Working directly under him were Cockburn, Inspector of Aeroplanes in charge of flight delivery, and Bagnall-Wild, Inspector of Engines. When the Air Battalion was formed in 1911, Fulton had been appointed to the No 1 Aeroplane Company at Larkhill. In 1912 he was one of the chief men at the Central Flying School and in 1913 the RFC Chief Inspector of Material, then he'd been given a free hand to start the AID and was very good at his job. We occupied two large hangars beside the Royal Aircraft Factory not far from the tree to which Cody used to tie his aeroplanes when testing his engines in the early days, the same principle as Roe's post. There were not many of us there, a total staff of about forty-five, but the department was expected to expand rapidly as we took over more inspection work.

The aeroplane was becoming deeply involved in the war and four of our men received the Distinguished Service Order (DSO), for daring aerial attacks, Samson, Collett, Marix and Spencer Grey, our first honours of the war. Samson had led the RNAS Air Reconnaissance flights to Belgium during August, during September an attempt to bomb the airship sheds at Dusseldorf and Cologne was unsuccessful, but during October Marix and Spencer Grey flying *Sopwith Tabloids* bombed the Dusseldorf sheds and Cologne Railway Station. The most famous attack was on the Friedrichshafen Zeppelin sheds and works by Lake Konstanz made during November 1914 causing serious damage. Sippe, one of our early AVRO pupils was on it with Babington and

Briggs, flying *Avro 504s*. These were important raids as Germany still believed in airships for war and Friedrichshafen was its main works. They were awarded the DSO and Distinguished Service Cross (DSC). H.G. Wells, the famous writer is reported to have said: '*The task that we are asking from our aviators is one of the most dazzling and terrible that men have ever faced.*' Around this time we learnt that Prince Maurice of Battenberg, brother of the Queen of Spain both of whom I'd had the pleasure of meeting during my second visit to Spain, died of wounds received in the first battle of Ypres. I was terribly sorry to hear this news.

Before 1914 came to an end, the Wright Brothers scored over the Government. C.G. Grey told us the weird details:

'But for the fact that everyone's attention is concentrated on the War, the announcement which appears hereunder would cause the biggest sensation that the aeroplane industry has enjoyed for quite a long time. For more than a year past the British Wright Company Limited, who own the original Wright Patent, had been suing the War Office in respect of infringement by the Forces of the Crown. One is now able to make it known that in order to relieve the War Department from unnecessary embarrassment during the stress of War, the British Wright Company made an offer to accept £15,000 in settlement of the original claim of £25,000 for the use of their patent. This offer has now been accepted. Obviously, after such an agreement had been reached, it is impossible for anyone who is not a millionaire to fight the Wright Brothers in court and, doubtless many people will jump to the conclusion that the industry is in for a very uncomfortable time in the future, owing to the royalties which will be claimed by the owners of the patent. However, I am informed by Mr Griffith Brewer, the well-known Patent Agent who operates for the British Wright Company, that the arrangement made with the Government covers all claims against all machines which are being built, or ever will be built, for either the Admiralty or the War Office. This includes not only those built to Government design but others ordered from firms who are building to their own designs, so that all Government contractors are exempt from any claim.'

Not a nice business!

Britain suffered no aerial attack until 19 January 1915, when Zeppelin airships dropped ten bombs on Yarmouth. Many did not explode but two people were killed and a few houses damaged. The airships followed the coast dropping further bombs, without doing much harm, as they made their way to Kings Lynn where seven fell and two more people were killed. Similar attacks followed at an average of two or three a month while the Press complained bitterly about the bombing of places which had no military significance. 'Wilful Murder... Murder by some agents of a Hostile Force... Death was due to injuries caused by a dastardly and illegitimate Act Of War.'

It was at the Department's suggestion that I should become a member of the RFC, the transfer was all done by paperwork and at Colonel Fulton's personal request, I remained attached to the civilian body of the AID. H.V. Roe looked me up and wanted me to return to AVROS. I also had a visit from A.V. Roe himself when he came to the AID sheds to have a look around. I gave him a little flight on a Sopwith. He wanted me back but I could not possibly consider leaving the AID now, I had become quite a fixture. There were only a handful of us testing machines, people came and went but I was the only regular one there. The same refusal applied when Tommy Sopwith approached me, asking me to come back since war production had been greatly stepped up. It was quite out of the question. The AID was doing a fine job, its creation was one of the best things that had happened in aviation. By thorough inspection and tests, the AID was making flying safe and I was quite content. Shortly after my transfer to the RFC, I was made a captain.

The man to be awarded the first VC of the war for the RFC was a man we knew well, Rhodes Moorhouse, who'd started his flying career working as a mechanic to Radley at Huntingdon. He had done quite a bit of flying, then gave it up when he married and, heir to a considerable fortune, the future looked rosy for him, but he didn't live to enjoy it. The award was given posthumously for courage shown in April 1915 when he descended to 300ft to bomb the railway at Courtrai in Belgium and was badly wounded. He died from his wounds and was buried in Dorset at his family's grounds in Netherby, in a small wood where he had planned to build a home for his family.

Another Victoria Cross was awarded to Reginald Warneford who distinguished himself by being the first pilot to bring down a Zeppelin. The airship exploded, the force of the explosion caused his monoplane to turn upside down, he righted the machine, but had to make a forced landing and was able to restart his engine and returned home safely. Warneford, a young man of 23 who'd learnt to fly at Hendon with Merriam only four months previously, had just been sent on active service and his mother at White

Lodge, Runfold, received numerous letters of congratulations but ten days later, he was dead. He'd not had sufficient flying experience. He'd tried to do a stunt at Buc on a large Farman while flying with an American journalist, the machine buckled and, fouled by its own propeller, the fatal crash occurred. A memorial was erected depicting a blazing Zeppelin in flight, a portrait of Warneford carved in stone, and three words, 'Courage. Initiative. Intrepidity.'

A report came from Paris at the beginning of September 1915 that Pégoud, the first man to loop, was killed by a German pilot returning his fire, then Colonel Fulton died in November. He had done a great deal for Government aviation and was sadly missed. We'd got on exceptionally well, fine tributes were paid to him. He had recently been appointed Assistant Director of Military Aeronautics and was in his office in the morning and feeling unwell, went to see a doctor who told him that an operation on his throat was necessary at once, but he died the same evening. Major Beatty became the head of the AID, an early military pilot of the Avro School to whom I'd once given instruction.

Up to now the pusher plane had been the popular machine for carrying or attaching guns since there was no propeller to get in the way of the fire, but they were inclined to be a slower type of machine, then Garros who'd been my Schneider competitor, came up with an idea which puzzled the Germans for a short time. He'd discovered a way of getting around the problem and was firing between the propeller of his tractor monoplane. What he'd done was simply to fit steel plates on the propeller which deflected the bullets that did not get through the turning blades, but then he was shot down and tried unsuccessfully to burn his machine before he was captured. The Germans, hitherto mystified by Garros's rapid fire from the front of a Morane, examined the machine and soon discovered the secret. Fokker, Dutch by birth and a naturalised German, introduced an interrupter gear on the Fokker monoplane, a monoplane which proved superior in speed and climb and deadlier with its new fast-firing device. Our losses were heavy from this point on.

Now there was a struggle for *Air Supremacy*... again and again the news was full of the 'Fokker Scourge'... We were in serious trouble and in January 1916, the RFC issued an order to its pilots.

> 'Until the Royal Flying Corps are in possession of a machine as good or better than the German Fokker, it seems that a change in the tactics employed becomes necessary. It is hoped very shortly to obtain a machine which will be able to successfully engage the Fokkers at present in use by the Germans. In the meanwhile, it must be laid

down as a hard and fast rule that a machine proceeding on Reconnaissance must be escorted by at least three other fighting machines. These machines must fly in close formation and a Reconnaissance should not be continued if any of the machines becomes detached. This should apply to both short and distant Reconnaissances. Aeroplanes proceeding on Photographic Duty any considerable distance east of the line should be similarly escorted. From recent experience it seems that the Germans are now employing their aeroplanes in groups of three or four, and these numbers are frequently encountered by our aeroplanes. Flying in Close Formation must be practised by all pilots.'

As Britain and her allies fought for supremacy during this critical period, we produced three types of synchronising gears but not until August 1916 did we make any real headway when George Constantinescu, a Roumanian and friend of Coanda, invented an efficient one. His synchronising gear replaced all other types which were being tried out, but 1916 truly was a very black year for the RFC. A lot of our men who were sent over the German lines soon after they qualified, were only half-trained. Many had short lives and no sound flying experience to carry them through. One little story reached us of a certain Lieutenant Thaye of the RFC who had been engaged in air combat over the German lines. With his machine badly damaged, his instruments not working, he ran into mist then came down giving himself up to a resident. 'Is this Holland?' 'No, Herne Bay.'

A Zeppelin had been brought down in the Thames Estuary by ground fire, then another airship was brought down at Cuffley, Hertfordshire, in September 1916 after nearly two years of Zeppelin raids. The public went wild. Several hundreds of people had been killed or injured as the result of airship invasions on England and the man who did it was Leefe Robinson who had used the new incendiary bombs especially designed for attacks on airships. The whole countryside was lit up as huge quantities of gas escaped, and the 33-ton airship slowly sank to the ground in flames, a burning mass seen 10 miles away in London. *What a death!* There were sixteen victims. The relics of the airship were put on show and people queued for admission. Mrs J.B. Kitson, owner of Cuffley Hill, presented the site on which the airships fell to the *Daily Express* on condition that a monument would be erected. This was duly done. Leefe Robinson was awarded the Victoria Cross and made a national hero, but that was not all. Money awards of several hundreds of pounds had been offered by private individuals since the first Zeppelin raids, to the first pilot to bring down a German airship. Leefe

Robinson received them all – £2,000, £1,000, £500, £500 – maybe more. Then the question of monetary awards to members of the RFC arose and as the outcome of this an Army Order was issued. 'Officers, warrant officers, non-commissioned officers and men are forbidden to accept presents in money from public bodies or private individuals in recognition of services rendered in the performance of their duty.' Shortly afterwards, Leefe Robinson was brought down in Germany and taken prisoner.

Within weeks of the Cuffley disaster, we had brought down two more airships with complete loss of German lives and it became obvious to the Germans that airships could not compete against the aeroplane. Six months later, in March 1917, Count von Zeppelin died at the age of seventy-seven. The Count had sunk all his money into his airships since the beginning of the century and had, despite tempting offers, refused to sell any of them to foreign powers, but over thirty had now been lost in the war, crippled by bad weather, bombed in their sheds, brought down by aeroplanes or by ground fire. The Friedrichshafen works which made most of the Zeppelins ceased production and turned to aeroplane construction. The following tribute was paid to the Count, 'Those who respect the truly patriotic doctrine, my country, right or wrong, would recognise the Count von Zeppelin as a great patriot albeit an enemy. Setting aside the fact he was an enemy, and a potent enemy, of this country, he stands forth as one of the dominating figures in the history of aerial navigation, and it is as such that he deserves primarily to be considered.' He was buried at the Pragfriedhof Cemetery, Stuttgart.

It was gone! The Blue Bird Café... our famous café at Brooklands was gone where friends had relaxed together and talked *Aviation*... It was reported that, 'a row of hangars and a canteen in the Brooklands aerodrome were destroyed by fire in the early morning 28 March 1917, the outbreak having, it is supposed, originated in the canteen. All aeroplanes are believed to have been saved. The local fire brigade prevented the fire from reaching the large workshops and offices. When one remembers how one smoked, and ran brazing lamps, and even engines and used plain oil lamps in the old sheds at Brooklands, regardless of open cans of petrol and even pans full of petrol for the washing of parts, it seems a miracle that the whole place was not burned out years ago. And now the poor old Blue Bird disappears in flames under a strictly regulated military control. So passes the scene of what were the happiest days in the lives of many of the old hands of aviation!'

Bloody April 1917 and 717 planes were brought down on the Western Front during that month. To think we'd all thought the war, now in its third year, would not last six months. Losses had never reached such a tremendous figure in one month of fighting since war began. There was a Royal Flying Corps verse going around, which created a vivid picture of our men at the Front.

> The young aviator lay dying,
> And as 'neath the wreckage he lay,
> To his comrades assembled around him,
> These parting words did he say,
> Take the Pistons out of my kidneys,
> Take the con rods out of my brain,
> From the small of my back take the crankshaft,
> And assemble the engine again!

I received the occasional Daily Routine Order to take machines to France and the following refers to a de Havilland I delivered at the beginning of April. I wonder what happened to it. 'Type de Havilland 4, No A/2166. To Officer Commanding, No 1 Aircraft Depot, St Omer. The bearer is delivering the above machine to the Expeditionary Force. He is to return to South Farnborough immediately on completion of this duty. Major Cox, The Officer Commanding Ferry Pilots, Royal Flying Corps, 3 April 1917.'

Shortly after this delivery I was asked jointly by the War Office and the India Office to fly a machine during a presentation ceremony in Leeds, a FE2b biplane built by Blackburns of Leeds and paid for by the people of Leeds. It was to be presented to the Indian Government for their Flying Corps. So I went to the Blackburn Works to collect it, and a facsimile of the Star of India was handed to me to be fixed on the front. It was a small brass plaque about four inches in diameter and I expressed that I should like to have one. With some difficulty, I did acquire one, they were inscribed, 'Heaven's Light, Our Guide.' I pocketed the plaques and flew off to Blackburn's Military Aerodrome at Roundhay where the ceremony was to take place. A great interest was taken in the Leeds plane. The Press reported the occasion as did the Blackburn's in-house magazine, *The Olympian*. 'The prospect of seeing a real battle plane attracted some 50,000 spectators to our aerodrome at Roundhay.'

During a luncheon at the Queen's Hotel on the day of the ceremony, Saturday, 15 April 1917, the Lord Mayor of Leeds welcomed his guests, General Sir David Henderson, head of the Royal Flying Corps, Lord Desborough, Chairman of the Imperial Air Fleet Committee and head of the

Volunteer Force, Lord Islington, Under Secretary of State for India and Mr J.E. Bedford, President of the Chamber of Commerce. Lord Desborough spoke, 'The Loyalty of the people of India during the war, from the highest to the lowest, is the best tribute that has ever been paid to the beneficence of British rule.' And Lord Islington, on referring to the Leeds plane, 'The gift of the aeroplane from the Leeds Chamber of Commerce will be deeply appreciated by the people of India.' General Henderson spoke of the need for fortitude in the present critical days, and of the RFC he said, 'Perhaps I have been in control of the RFC longer than any man should have such a job, but in that time I have learnt to know the officers and the men of the Corps, and I tell you that these young men are the very salt of the earth.'

After the luncheon we made our way to the aerodrome which had been roped off for the occasion. A body of Special Police, on foot and mounted, a guard of honour and a squadron of the RFC from a nearby station, were present. As a curtain raiser, a Naval airship passed over at low altitude. It was a great day for the Blackburn workers who seemed pleased that I should be flying their machine, giving me a lengthy write-up in their magazine:

'We are fortunate in having to pilot the machine a man of the eminence and long record of Captain Pixton, who first attracted notice by his plucky attempt in the *Daily Mail* Circuit of Britain in 1911, which ended in disaster at Spofforth close by. How well we remember his arrival at Harrogate, lame and shaken by his contact with dear Mother Earth, and the excited applause of the surging crowd on the Stray. Pixton, the hero of that hot summer day, was certainly the hero of this clear spring afternoon. He looked very little older, and in his smart military uniform a great credit to the RFC, which he has adorned for three years. Attached to a Government factory, Pixton had the luck and good management, to be able to claim a noteworthy freedom from even the simplest mishap, and his performances must have contributed to the nation's success in the air.'

I stood by the machine as the Lady Mayoress broke a bottle of champagne over its nose, only half full because of the war need of avoiding waste, 'I name this battle plane *Leeds*. May it have the good luck we all desire for it, and may Heaven guide its pilot through all dangers.' The facsimile was handed to her and she lightly attached it to the front. Lord Desborough stepped forward. 'I ask Lord Islington to accept this gift from the Leeds Chamber of Commerce, for the Indian Government.' Lord Islington, in accepting the gift, said 'For all time to come, *Leeds* will be brought into intimate association with what will be regarded in the future as one of the most indispensable parts of the defence of India, namely, the air system. In accepting this gift for the Indian Government, I place it in the charge of Sir David Henderson, Commander of the Royal Flying Corps, for use at the Western Front.' And so Sir David accepted it. 'I give two undertakings, that it should be used in France, and that when the war is over, this aeroplane, or its successor, by the name of Leeds should be handed over to the Government of India as the gift of the Leeds Chamber of Commerce.'

The Lady Mayoress, Lord Desborough, Sir David Henderson, Mr Bedford, myself as the pilot and a few others, each were presented with a mahogany box containing a memento of the occasion in the form of a full-size silk Indian flag. I then got the machine started, gave a short exhibition, and made a second flight over the cheering crowds with Lord Desborough as my passenger. That evening, the chairman of Blackburns invited me to a dinner at the Queen's Hotel where I congratulated the firm on their enterprise and their high standard of workmanship. Next day at noon I left with a colleague, Lieutenant Vickers of the AID, flying the Leeds plane to Farnborough from where it would leave for the Western Front, and Blackburn's *Olympian* ended their report of the occasion, 'Sweetly she rose, higher and higher, steady as a seagull, and the last we saw of our handiwork was a speck which a large, fleecy cloud swallowed up.'

During my spell at Farnborough, a most unusual incident happened to me concerning an Irish Lieutenant. I'd sold the Model T Ford car I'd bought in Southampton, replacing it with a Buick. One day when I was taking away a gallon of petrol from the AID sheds for it, the lieutenant who was not particularly friendly towards me, spotted me and promptly aimed his revolver, firing shots at the can which immediately started to spurt petrol. With my fingers in the bullet holes, I ran as fast as I could leaving a trail of petrol behind me, to empty what was left of the precious essence which was very hard to come by, into the car tank. There was a little left, but had I not been nippy on my feet, the can would have surely emptied, and as I screwed on the cap I shouted over to him, '*Thank God you're a good shot.*' What confidence, and what nerve!

But the nearest I'd come to death was when an Avro 504 nearly got me. It was undergoing tests and the engine would not start. Everyone was having a pull at the propeller and I offered to lend a hand. The usual thing was to have the engine off, turn the propeller, then switch on, but try as I might this one just wouldn't start. 'Switch the engine on as I swing it,' but the propeller suddenly started when I wasn't ready catching me off balance. Usually those who have overbalanced have fallen into the turning propeller, being either killed or shockingly mauled. I was lucky. Since this kind of accident was occurring too frequently, we'd made a slight change in the orders. Instead of shouting to the pilot 'on' and 'off' which sounded too much alike, we called 'contact' and 'off'. And a switch was nowadays put on the outside of machines which could be switched on by a second mechanic who was in more of a position to see the man who swung the propeller. The French single-seater SPAD made chiefly of plywood, almost got me too. I was gathering up speed on take-off when, on the point of rising, the wheels seized up and all the spokes splayed out. It had the makings of a serious accident but the SPAD turned over crashing at the side of the runway and I walked away with only a few bruises. The wheels had not been properly oiled.

The AID closed at Farnborough during 1917 as a testing area. Instead of having the one testing place, they introduced inspection centres all over the country called Acceptance Parks, necessitated by increased production of machines for the war. In charge of them was Conway Jenkins, ex-Avro pupil and first to crash the *Pixie Plane* of 1911. During December 1917, I went to the Acceptance Park at Newcastle for a period of six months as nearby was the Vickers-Armstrong works whose machines I tested and flew to Farnborough for despatch to the Front. I was a bad navigator. I flew by guesswork, or by instinct more than anything, first plotting the course on the ground with a map, then rising above the clouds judged the wind direction and made allowances for it, set the course, then fly for a certain time and was always around the right spot on descending. More often than not I used to get up to 7,000ft where it was generally nice and calm. As I headed for Farnborough one day, I saw a circular rainbow, a beautiful phenomenon, one of three I'd seen when flying high and in line with the sun and rain, but how surprisingly few pilots had seen one. There was another thing I used to mention, static electricity running up and down the plane as I flew in thunderstorms. I was not believed. Neither was I believed by many of my friends when I told them that I had seen sparks like a spider's web shoot out in front while flying through black clouds. This had occurred several times but I suppose I'd done more flying than most pilots and had seen more.

On 1 April 1918, April Fool's Day, air power was united in one body. The Royal Naval Air Service and the Royal Flying Corps were merged into **The Royal Air Force**. The amalgamation was not popular due to internal military and naval squabbles, but as its teething problems were resolved, it proved to be the finest thing to happen within the services. Sir Hugh Trenchard headed the new service as Chief of the Air Staff, but resigned within days of the announcement, and accepted command of an independent force created solely for offensive purposes, free to operate on its own.

However, he later accepted the position and did much to strengthen the Royal Air Force as a single service. As RAF already stood for Royal Aircraft Factory, something had to be done to avoid confusion, so the factory was renamed 'The Royal Aircraft Establishment'. There was also the question of uniform. The new blue uniform of the Royal Air Force was not liked by many, but the men were weaned onto the new blue, continuing to wear the RFC khaki or the RNAS blue at certain times, but not at others. Women played a part in the war and those of the Women's Auxiliary Corps, the Women's Royal Naval Service, and the Women's Legion of Motor Drivers, were invited to enrol in **The Women's Royal Air Force** which comprised four main sections – Technical, General, Clerical and Domestic.

After I'd been at Newcastle for six months, I was posted to Dublin for a further six months as an Inspector of Aerodromes and Landing Grounds. I carried the following authorisation, 'H.Q. Royal Air Force, To All Commanding Officers, Royal Air Force, Ireland. The bearer, Captain C.H. Pixton, Royal Air Force, is on Special Duty looking for suitable Aerodromes and Emergency Landing Grounds. Please assist him as far as possible and place such transport as may be available at his disposal in order to carry out these duties, 27 May 1918.' I was then transferred to the AID Headquarters in London at Clements Inn during October 1918, where my responsibilities were to visit aircraft factories and report on them.

So far we had not made use of the parachute in the war although successful tests had been made with the Guardian Angel Parachute. Its inventor, E.R. Calthrop, believed that all pilots should be issued with them, none had failed to open during tests, no one had been injured or killed, but Government officials, though expressing their admiration, failed to adopt them officially. Letters flowed in to the Press.

> 'If the use of the parachute were introduced for airmen, the probability would be that an aviator would be tempted to use it in cases where he may apparently have lost control of the machine, and could by perseverance right it, using the parachute as a means of saving his life and at the same time causing the loss of the machine.'

Not a popular thought. 'Human life is more valuable than aeroplanes, and the sacrifice of a few aeroplanes behaving as death traps, is a small price to pay for the saving of as many human lives by the comparative certainty of the Guardian Angel Parachute.'

The Aeroplane had reported, 'On 13 June 1918, Major Baird, Parliamentary Air Secretary, admitted to the House and the world at large that experiments had been carried out with the dropping of parachutes from aeroplanes, but implied that no practical method had yet been devised for their use.' The appeal that parachutes should form part of an airman's equipment was intensified and, at last, on 30 October 1918, a statement that certain machines at the Front would be equipped with them was made but, twelve days later, on 11 November 1918...

War was over.

Garros, with another pilot, had escaped from a German camp after three years imprisonment since being shot down in his plane with metal plates on its propeller, but as soon as he was back he requested to be attached to the flying squadron at the Front once again. The headlines had been, 'Garros and Marechal Free... Honours for Garros and Marechal.' Then was reported 'The disappearance of Roland Garros,' followed by 'Lieutenant Garros was shot down and killed on 5 October.' Had he waited another few weeks there would have been no fighting, no need to return to the Front. I had very pleasant memories of him in Monte Carlo, a fine pilot with a tragic end. This was echoed throughout the Press. 'Amid the millions of tragedies of the war, his death stands out as one of the most tragic and most heroic.'

The king's message to the Royal Air Force at the end of the war contained, 'The birth of the Royal Air Force, with its wonderful expansion and development, will ever remain one the most remarkable achievements of the Great War.' But thousands were dead. Many star pilots, British, French and German alike, had disappeared from the scene. Many of my friends and many I knew slightly, had gone, some in air combat and some by other causes.

To mention a few: Victor Mahl, following an operation for appendicitis; Colin Pizey, of dysentery while on a mission to Greece; Sir George White of Bristol's at the age of 62; Cedric Lee, inventor of the Circular plane, during service, Eric Pashley, one of the Pashley brothers at Brooklands, in action; Frank Cody, Cody's youngest son, in air combat; John Petre, another of the Petre brothers, in action; Jimmy Valentine, while on a mission to Russia, after an operation; Harold Barnwell, at Vickers' aerodrome in Kent, failed to come out of a spin; Ian Henderson, Sir David Henderson's only son; Gordon Bell, in France, crashed during a test flight...

Among those wounded were Reynolds, who lost the use of his left hand and Robert Loraine, wounded in the lung and leg but able to return to the stage. Victor Sassoon, badly wounded in the legs, took to owning horses. Sassoon received a total disablement pension which he gave to C.G. Grey to form a fund for pilots in difficulties.

Then scarcely had the Armistice been established when a flu' epidemic struck the country. The losses were heavy and included Hucks who had gone through the war unharmed. Captain Wood promoted to major, died of meningitis produced by an infection of the throat following flu'; Henry Maitland died at Cherbourg of pneumonia following flu' and also Leefe Robinson, famous for bringing down the airship at Cuffley, who had been a prisoner of war.

On a brighter side some of our colleagues had married and included H.V. Roe to Marie Stopes, Fellow and Lecturer in Palaeobotany at University College London. The marriage took place at St Margaret's Church, Westminster. H.V invited me but I couldn't get away on that day... Grahame-White had remarried. His wife was starring with Harry Lauder in *Three Cheers* at the Shaftsbury Theatre... Herbert Spottiswoode, the well-known figure at Brooklands in the early days and friend, who had been touring Germany at the outbreak of war and was taken prisoner but released shortly afterwards, had married... Prince Serge de Bolotoff, remembered at Brooklands for his triplane which never flew, married the daughter of Harry Selfridge, of Selfridges, Oxford Street...

I was still with the AID as 1918 came to an end and one of my last duties with them was to hand over an AID machine to Lord Weir, Secretary of State for Air. Lord Weir, accompanied by the Mayor and Lady Mayoress of Manchester, arrived early on the morning of 20 December at the National Aerodrome Factory, Heaton Chapel, Manchester, where the presentation was to take place. Brigadier General Bagnall-Wild of the AID received our guest and presented AID personnel to him before commencing on a tour of the works. Those presented included Rear Admiral W.J. Anstey, Assistant to Controller, Lieutenant Colonel H.W. Outram, Chief Inspector of Aeroplanes and Major G.P. Bulman, Deputy Chief Inspector of Engines. We gathered round the presentation aeroplane, a de Havilland DH9, which was given to Lord Weir on behalf of the staff of the AID and christened AID by the Lady Mayoress. After, I gave Lord Weir a demonstration flight and, thus done, we went to the town hall for luncheon.

Since I first stepped into Brooklands in June 1910 to 1918 and after 1918, I had flown or test-flown many planes of varying shapes, sizes and engine power, about eighty types in total flying about 3,500 hours. I didn't do much flying after 1918, but those I'd flown before and during the war included:

- **Avro** Triplanes, Avro Type D biplane, Avro School Farman Boxkite, Avro 504

- **Bristol** Boxkite, Bristol Tabateau biplane, Bristol England biplane, Bristol Prier monoplane, Bristol Coanda monoplane, Bristol Coanda biplane, Bristol Burney Hydro, Bristol Scout
- **Sopwith** pusher Hydro, Sopwith tractor Hydro, Sopwith *Daily Mail* Circuit Hydro, Sopwith 3-seat and 2-seat Standard, Sopwith Tabloid Scouts, the Sopwith Schneider Tabloid Floatplane, Sopwith Bat Boats, Sopwith Torpedo Hydro, Sopwith Pup, Sopwith Camel, Sopwith 1½ Strutter, Sopwith Triplane
- **Royal Aircraft Factory** BE 2a, 2b, 2c, 2d, 2e, 12, FE 2b, 2d, 8, RE 5, 7, 8, SE 5
- **de Havilland** 4, 5, 9
- **Blériot** 2-seat and Blériot Parasol 1-seat
- **Caudron**
- **Henry Farman** HF20
- **Maurice Farman** S7 Longhorn and S11 Shorthorn
- **Martinsyde** G100 and G102 Elephant
- **SPAD**
- **Armstrong Whitworth** FK8
- **Vickers** Gun Bus, FB5 and FB 14T. (T-type tail)

There were a few I'd attempted to fly in the early days of Brooklands, the *Hammond triplane*, the *Hawkins triplane* at Winchester and the *Martinsyde monoplane* with a 30hp JAP engine. Also there were the special ones I'd flown, the *Sopwith Tweenie* for Churchill, the delivery and presentation of the *Blackburn Leeds* plane for India and the *DH9 AID* plane for Lord Weir.

The types of engine (horsepower) fitted to the planes I'd flown included the 30 JAP, the 35 and 100 Greens, the Empress Rotary, the 50, 70, 80 and 100 Gnomes, the 60 ENV, the 70 Renault, the 50, 120 and 130 Clergets, the 80 and 100 Le Rhones, the 100 Anzani, the 90 and 150 RAF, the 120 and 160 Beardmores, the 200 Puma, the 200 Salmson, the 225 Sunbeam, and the 250 Rolls.

The war had been fought mainly with biplanes since the RFC monoplane ban of 1912 when manufacturers had turned to biplane production and had continued to do so even after the ban had been removed.

Flying would never be the same again. It had grown from child to adult overnight and a new type of people had appeared on the scene, different from the closely-knit crowd of the old days. Hendon was going to open next season, but parts of the track at Brooklands had been pulled up and it would be some time before it could be used again. Aviation had advanced more than it would have done in twice or thrice as many years of peace conditions and beside the small scouts of 1914, there stood huge multi-engine machines with great changes in performance, for example, altitude before war 7,000ft, after 30,000ft; speed before war 70 mph, after 150 mph. And no longer did Britain lag behind in Aero Engine Manufacture. Royce of Rolls Royce, who had not been very interested in flying while Rolls was alive, had set to with enthusiasm during the early days of the war designing and producing the much needed high-power British engine.

With the war over and with bigger machines, there was talk of operating *daily passenger services*. Handley Page proved the likelihood of such services in a practical fashion when a crowd went up in one of his machines, a standard bomber powered by four Rolls Royce engines, for a flight lasting thirty-three minutes, a world record. This is what was said about it. 'Forty-one up and room for more. On 15 November 1918, a world record was created for the largest number of passengers ever carried in an aeroplane when Clifford Prodger, an American pilot, took a Handley Page four engine super giant biplane with forty passengers on board for a cruise over London at a height of 6,500ft. In addition to this load, petrol was carried for a six-hour flight, so that these forty passengers could easily have been carried across the Channel to Paris had circumstances permitted.' It was referred to as '*A new era in Aviation*', but it took several years before passenger services became a regular and reliable means of travel.

1910 – 1918
DEVELOPMENT OF THE AEROPLANE
Progress by Usage

1910	Experimental, Challenges, Races.
1914	Reconnaissance
1915	Fighters
1917	Light Bombers
1918	Heavy Bombers
Later	Civil Aviation, Light Planes, Long Flights, Clubs.
Altitude	Before the War 7,000ft, after 30,000ft. (approx)
Speed	Before the War 70mph, after 150mph. (approx)

There would never be another Brooklands. *Brooklands*, where we'd tucked ourselves away in one small area with our aeroplanes to escape public criticism and prejudice... the nerve centre of pioneer aviation... the hub of British aviation. Here I had spent the happiest days of my life. Here I had seen an aeroplane leave the ground for the first time, Roe on his Triplane! I was lucky to be at the right place where the first aeroplanes were being made, and those early pioneers of Brooklands were the finest fellows it has ever been my good fortune to meet – quiet, unassuming and kind hearted – who flew for the sheer love of flying with very little reward, never thinking they were doing anything out of the ordinary, never thinking they were making history, but these were the men who handed on the 'know-how' to later pilots, and looking back, there are two men, two men only, who stand out above the rest.

They are the two men who made the first flights in this country at about the same time. The greatest of them all, A.V. Roe and S.F. Cody, whom I looked upon as my two greatest and closest friends, they are the ones in Britain who discovered the hard way the basic principles of flight by their own failures and successes – a small part of history, but the most dominant technological phase of the twentieth century.

Stella
Howard Pixton's Daughter.

My father retired to the Isle of Man in 1932. He died in his 87th year, in February 1972, and is buried at Jurby Churchyard. I was born during his retirement... It was said of father:

'Pixton could fly any machine.'

Company Advertisements
Including those that use Howard Pixton's name for promotional purposes.

EVERYTHING FOR AEROPLANES

Whatever you may want for an aeroplane, be it bolts, screws, wire, or wheels, can be obtained from us at the lowest price, consistent with quality.

ENGINES, PROPELLERS, or COMPLETE MACHINES
new and second-hand.

Complete machines or parts made to inventors own designs, of guaranteed workmanship and finish.

Sole makers of the "AVRO" PLANE, the safest machine and easiest to handle.

Passenger flights arranged on own Triplane, genuine Farman or Bleriot.

FLYING TAUGHT, TILL FULL CERTIFICATE IS OBTAINED, FOR £50 INCLUSIVE.
FLYING GROUND—WEYBRIDGE.

A.V. ROE & CO., THE AVIATORS' STOREHOUSE, Brownsfield Mills, MANCHESTER.

1910

GREEN'S ALL BRITISH AERO ENGINES

BROOKLANDS EASTER MEETING, 1911.

PIXTON flying the ROE BIPLANE fitted with

30/35 H.P. "GREEN" ENGINE

flew for 87 mins. 22 secs. in a very strong wind.

This was the BEST FLIGHT of the day, and won the BROOKLANDS ENDURANCE COMPETITION, creating at the same time a RECORD for 30-35 h.p. engines.

The Engine was identical with the one which scored so remarkable a success in the PATRICK ALEXANDER COMPETITION.

THE GREEN ENGINE COMPANY (Sole Licensees under the Green's Motor Patents Syndicate, Ltd.), **55, Berners St., Oxford St., London, W.**

Telegrams: "Airengine, London"

1911

Advertisements

Bristol Advertisement

"BRISTOL" STANDS FOR RELIABILITY AND STRENGTH

This Photograph shows an
All British "BRISTOL" Biplane
(Flown by Mr. Pixton)

Flying in a gusty wind of 30 m.p.h. at Brooklands, June 24th. This was the **ONLY MACHINE IN ENGLAND FLOWN ON THAT DAY.**

'Bristol' Aeroplanes are Reliable & Strong
They are BRITISH BUILT from the Finest Materials.
For full Particulars write to the Builders.

The... **"BRISTOL" FLYING SCHOOLS** are at **Salisbury Plain** and **Brooklands.**

THEY ARE EQUIPPED WITH **"BRISTOL" BIPLANES,** FITTED WITH **"GNÔME" ENGINES.**

Expert Tuition ensuring Rapid Proficiency.

SPECIAL TERMS TO NAVY & ARMY OFFICERS.

THE BRITISH & COLONIAL AEROPLANE CO. LTD.
BRISTOL, ENGLAND
TELEGRAMS "Aviation" Bristol
TELEPHONES No 2 P.O. 74 } National 3074

1911

E.N.V. Advertisement

E. N. V.
ALL BRITISH AVIATION MOTORS.
MANVILLE PRIZE of £500

Brooklands Racing Season Aggregate Prize of **£150**

Brooklands Duration Prize, Oct. 4th. **£30**

All Won by C. Howard Pixton, Esq., using a 60 h.p.

E. N. V.

Address all enquiries to—
The E. N. V. Motor Syndicate, Ltd.,
4, Hythe Road, Willesden, London, N.W.

1911

Advertisements

April 30, 1914. *The Aeroplane.*

THE SCHNEIDER CUP RACE
WAS WON ON A
SOPWITH BIPLANE
Piloted by MR. C. HOWARD PIXTON.

Maximum Speed over 10 kms.
(Closed circuit with four corners)
92·1
miles per hour
(**148·3** kms. per hour).

Total Distance of Race
150 sea miles (172 miles) in
2 hrs. 13 secs.

The Sopwith Biplane starting in the Schneider Cup Race

300 KILOMETRE SEAPLANE RECORD 2 hours, 9 mins., 10 secs.
The Sopwith proved itself to be the fastest seaplane in the world.

THE SPEED-RANGE OF THE LATEST SOPWITH SCOUT was timed officially at Farnborough as from 94·9 to 39·6 miles per hour.

PREVIOUS RECORDS on the preceding type of **SOPWITH BIPLANE** (80 h.p. Gnôme).

British Height Record (Pilot alone) 11,450 ft.
British Height Record (Pilot and 1 Passenger) 12,900 ft.
British Height Record (Pilot and 2 Passengers) .. 10,600 ft.
WORLD'S Height Record (Pilot and 3 Passengers) 8,400 ft.
ALSO, British Duration Record, 8 hrs. 23 mins. with British A.B.C. Engine 40 h.p.

The Sopwith Biplane flying in the Schneider Cup Race

THE SOPWITH AVIATION CO.
(CONTRACTORS TO HIS MAJESTY'S ADMIRALTY AND WAR OFFICE)
Offices and Works: KINGSTON-ON-THAMES.

Telephone: 1777 Kingston. Telegrams: "Sopwith, Kingston."

1914

Advertisements

THE SCHNEIDER CUP RACE

RESULT

1st—PIXTON ON SOPWITH BIPLANE

100-H.P. MONOSOUPAPE GNOME MOTOR

Beating all Speed Records for Hydro-Aeroplanes up to 300 Kms.

2nd—BURRI ON F.B.A. FLYING BOAT

100-H.P. MONOSOUPAPE GNOME MOTOR

For Particulars concerning the new Monosoupape Gnome

APPLY TO THE

GNOME ENGINE COMPANY.

(SOCIÉTÉ DES MOTEURS GNOME.)

1914

Advertisements

PIONEERS OF MODERN AVIATION

C. HOWARD PIXTON had his first aeronautical experience with A. V. Roe at Brooklands in June, 1910, and later in the same year he went over to the Boston meeting. In 1911 he entered for the Brooklands–Brighton race, but did not finish. In April, 1914, he won the Schneider Cup, the year after Maurice Prevost. Shell was already hard at work preparing for the Shell Aviation Service, which was first established in 1919. The system of ground organisation is now so efficient that during a recent record flight to the Cape the pilot was able to stop, obtain supplies of Shell Aviation Petrol, and be off again within 20 minutes.

AEROSHELL LUBRICATING OIL · SHELL AVIATION PETROL · SHELL CARNET

YOU CAN BE SURE OF SHELL

Later...

Aero Club of Great Britain

Royal Aero Club of Great Britain

Royal Aero Club
OF THE UNITED KINGDOM

This is to Certify that C. H. Pixton, Esq of Durrington, Amesbury, has been duly entered upon the Competitors' Register for 1912, and is entitled to take part in any Competition held under the Competition Rules of The Club.

This Certificate expires on December 31st, 1912.

Harold E. Perrin
Secretary

Register No. 95

THIS PERMIT MUST BE EXHIBITED AT ALL COMPETITIONS WHEN REQUIRED.

Royal Aero Club
OF THE UNITED KINGDOM,
166, PICCADILLY LONDON, W.

No. 160

1913

This is to Certify that C. Howard Pixton of ROYAL AERO CLUB, has been duly entered upon the Competitors' Register for 1913, and is entitled to take part in any Competition held under the Competition Rules of The Club.

This Certificate expires on December 31st, 1913.

Harold E. Perrin
Secretary

THIS PERMIT MUST BE EXHIBITED AT ALL COMPETITIONS WHEN REQUIRED.

Royal Aero Club
OF THE UNITED KINGDOM,
166, PICCADILLY, LONDON, W.

No. 180

1914

This is to Certify that Mr. Cecil Howard Pixton of Brooklands, King's Road, Walton-on-Thames, has been duly entered upon the Competitors' Register for 1914, and is entitled to take part in any Competition held under the Competition Rules of The Club.

This Certificate expires on December 31st, 1914.

Licensed under Article 70 of the Regulation of the Fédération Aéronautique Internationale

Harold E. Perrin
Secretary

THIS PERMIT MUST BE EXHIBITED AT ALL COMPETITIONS WHEN REQUIRED.

Receipt and Telegrams

Received 24 June 1910
Deposit Thirty pounds (£30)
To be returned when lessons complete less cost of repairs to aeroplane.

H.V. Roe's receipt to Howard, 1910.

POST OFFICE TELEGRAPHS.

Aldershot

TO Pixton Aerodrome Brooklands Weybridge

Hearty congratulations I'm late nevertheless sincere

Cody

Cody's telegram referring to Howard winning the Manville Prize, 1911.

POST OFFICE TELEGRAPHS.

Wien 16th 4.5pm

TO Pixton Pilote Amesbury

Mes meilleurs félicitations General Coanda

General Coanda's telegram referring to Howard's wind-flying World Record on a Bristol Coanda Monoplane, 1912.

Letters

25th May 1911.

- Pixton Esq.,
   Brooklands Flying Ground,
      Weybridge.

Dear Mr Pixton,

It would afford Mr Lee and myself very great pleasure if you could dine with us at the House of Commons on Tuesday next at 8 o'clock to meet a few of the Members who are interested in aviation.

Yours faithfully,

Arthur du Cros

The British Government begins to take an active interest in flight, 1911.

THE ROYAL AERO CLUB
—— OF THE ——
UNITED KINGDOM,
166, PICCADILLY,
LONDON, W.

11th October 1911

Dear Sir,

MANVILLE £500 PRIZE

My Committee at its meeting last evening, examined the certificates of recorded flights in the above Competition, and I have pleasure in informing you that the prize of £500 offered by Mr. E. Manville was unanimously awarded to you.

Yours faithfully,

Harold E. Perrin
Secretary.

C. H. Pixton Esq.,
   c/o British & Colonial Aeroplane Co.,
      Filton House,
         Bristol.

Confirmation Howard wins the Manville Prize, 1911.

On Winning the Schneider Trophy

Howard's postcard to his mother reads:

'Of course you know all about the race by now. Our machine simply ran away from everyone. The picture's taken just upon landing after the flight and Mr Schneider, the giver of the "cup", is on my left. I am returning on Wednesday evening and home again Friday morning.' 1914.

Howard's joyful win!

To the casino!

Seven Days After!

BROOKLANDS AERODROME.

27th April, 1914.

Dear Sir,

FLYING TO THE DANGER OF THE PUBLIC AT BROOKLANDS.

With reference to my conversation with you yesterday afternoon, I beg to confirm what I then said to you regarding the dangerous nature of your flying, when you promised to exercise greater care in the future.

As you are doubtless aware, the Home Office have appointed, or contemplate appointing, special Inspectors to visit flying grounds and look out for dangerous flying, and if the management were to take no steps to put a stop to it, the Home Office would undoubtedly exercise their powers and close the ground to the public.

An acknowledgment of this letter will oblige,

Yours faithfully,

Frank Wright

Howard Pixton, Esq.,
"Brooklands," Kings Road,
    WALTON-ON-THAMES.

'A prophet is not without honour, but in his own country.'

War Time Passes

[Signature: E Howard Pixton]

The person whose signature appears above is employed by the Aeronautical Inspection Department and is therefore SERVING HIS COUNTRY.

[Signature] Lt Col
.... ~~Major.~~

Chief Inspector,
Aeronautical Inspection Dept.

Date 23.2.15

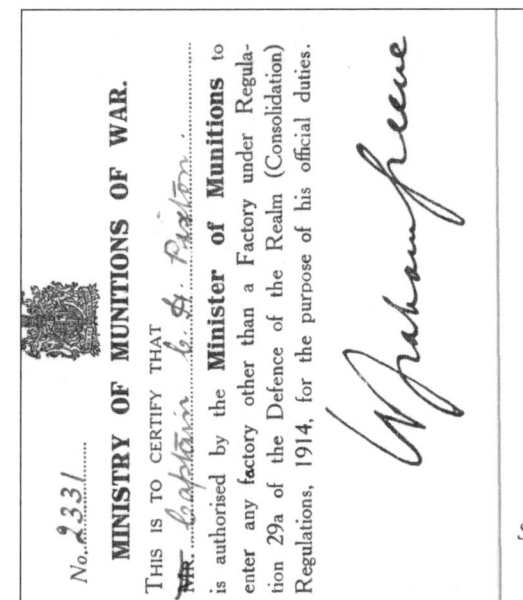

Gale & Polden, Ltd., Printers, Aldershot. 7,068-v.

No. 2331

MINISTRY OF MUNITIONS OF WAR.

THIS IS TO CERTIFY THAT ~~Mr.~~ Captain C. H. Pixton is authorised by the Minister of Munitions to enter any factory other than a Factory under Regulation 29a of the Defence of the Realm (Consolidation) Regulations, 1914, for the purpose of his official duties.

[Signature: Graham Greene]

PORTRAIT OF THE OFFICER.

Signature of the Officer— *[Signature: E Howard Pixton]* Capt. RAF

No. 2331

Eyre & Spottiswoode, Ltd., East Harding Street, E.C.4.

War Time Instructions

ROYAL FLYING CORPS.

...... Capt Pixton R.N. ...... has permission to enter and leave the town of ST. OMER on ... 27th & 28th Nov ...

                                                    F H ......
                                                        Major.
In the Field.   D.A.A. & Q.M.G.
27. 11. 1915.   Royal Flying Corps.

TYPE De Hav 4. NO A/2166. PILOT Captain Pixton

TO:-
    OFFICER COMMANDING,
        No.1.Aircraft Depot,
            ST. OMER.

        The bearer is delivering the above machine to the Expeditionary Force.

        He is to return to South Farnborough immediately on completion of this duty.

[Stamp: FARNBOROUGH O.C. FERRY PILOTS 13 APR 1917 FERRY PILOT'S OFFICE, S.A.D.]

                                            P H Cox
                                                    Major,
Date _____ 1917.   Officer Commanding Ferry Pilots,
                                                Royal Flying Corps.

For "forced landings" in England, please ring up "Regent 8,000, Ex.242." and also "42 North Camp".

                                                Headquarters,
                                                Royal Air Force.
To All Commanding Officers,
Royal Air Force Units,
I R E L A N D.

        The bearer, Capt.C.H. Pixton, Royal Air Force, is on special duty looking out for suitable Aerodromes and Emergency Landing Grounds. Please assist him as far as possible and place such transport as may be available at his disposal, in order to carry out these duties.

                                                    Major,
                            for O.C.,55th Wing,R.A.F.

DUBLIN.
27/5/18.
E.M.C.

After the War

Telephone No:
Regent 8000.
Telegraphic Address:
"Airministry, London".

AIR MINISTRY,
KINGSWAY,
LONDON, W.C.2.

C 96811   20 Apr 20

All letters on the undermentioned subject should be addressed to THE SECRETARY at the above address and should quote:

Air Ministry Refce. P.4.F.   Your Refce.

Subject :-

Sir,

    I am commanded by the Air Council to inform you that you have been placed on the Unemployed List of the Royal Air Force with effect from...... 11...6...19......
and you will cease to draw pay from Air Force funds from that date.

    I am to say that, on demobilisation, you will retain the rank of...... Capt. ...... but this does not confer the right to wear uniform, except when employed in a military capacity or on special occasions when attending ceremonials and entertainments of a military nature.

I am,
Sir,
Your obedient Servant,

C R Briptocke

Capt C H Pixton
Ye Retreat
Church Rd.
St. Annes on Sea

Pages from a log book

PAST EXPERIENCE—cont.

Types Flown.	Hours Flown on each Type.	No. of Accidents.	Remarks.
Over 80	not known	2 (Minor) 1 (Major)	

I CERTIFY that the information on this and the preceding pages is accurate to the best of my knowledge.

Date 14/7/30.   Signature of Pilot *Howard Dixon*

Howard flew over 80 types of planes, mainly during 1910–1918. He virtually gave up flying when the war ended.

PAST EXPERIENCE AS A PILOT.

MILITARY.

Date of Commission 1/4/15
" Qualified for Wings 1/4/15
" Demobilized 11/6/19.
Highest Rank held CAPTAIN – RFC. + RAF
Decorations Awarded —

Approx. Hours Flown { By Day 2000
                     { " Night NIL

Total Hours Flown on Service 2000

CIVIL.

Approx. Hours Flown { By Day 1500
                    { " Night NIL

Total Hours Flown as Civilian Pilot 1500

Approx.: Total Flying Time as a Pilot to date = 3,500.

The First Round Britain Race, 1911

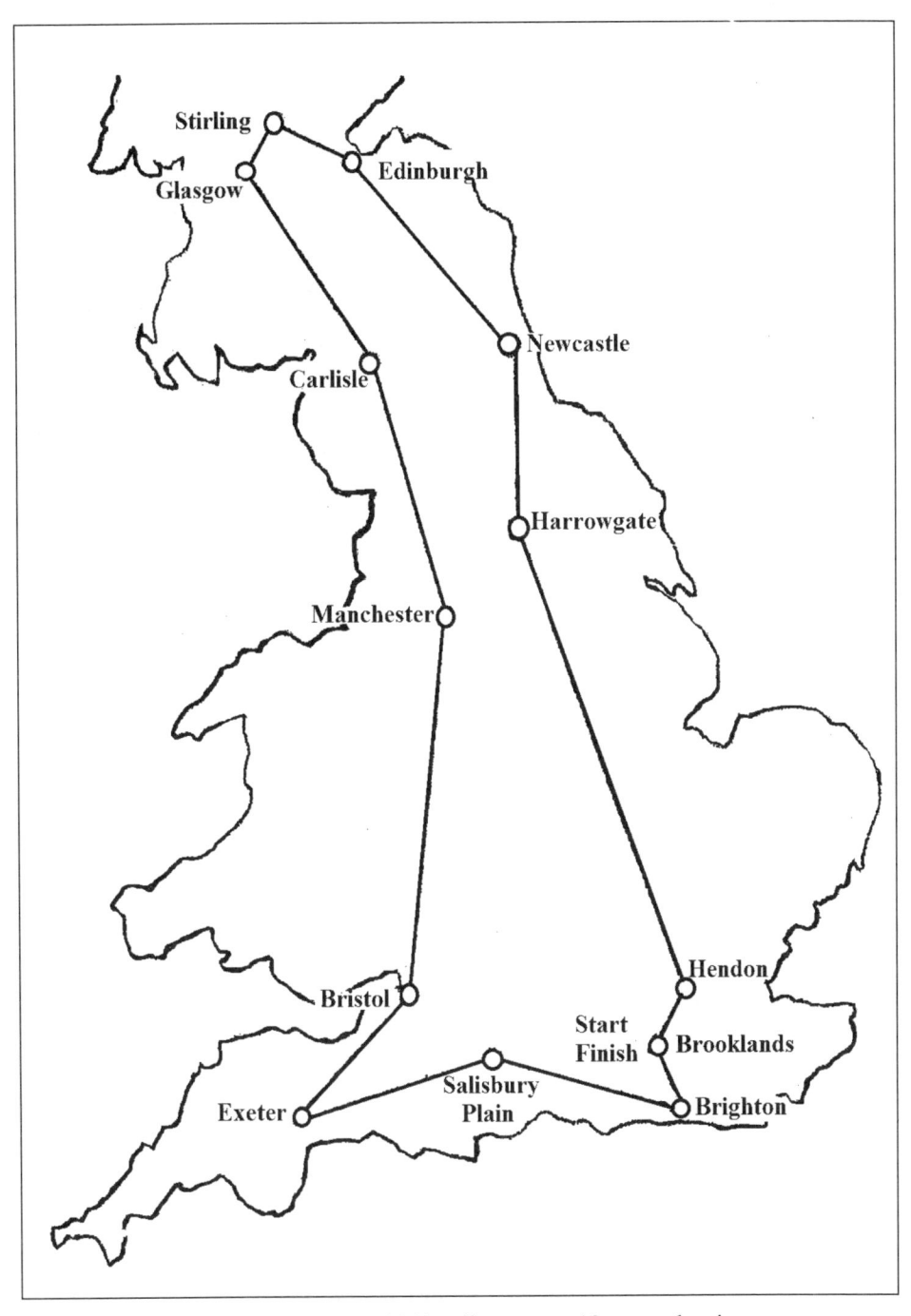

An Endurance Race, 1010 mile course, 12 control points.
First two places claimed by France.

The Tabloid

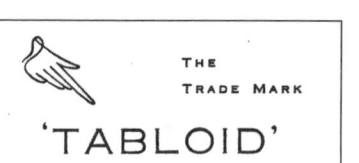

THE TRADE MARK
'TABLOID'

The word 'TABLOID' is a registered Trade Mark.

The word 'TABLOID' denotes a Burroughs Wellcome & Co. product, and is applied to many classes of products.

The word 'TABLOID' on a package, wherever purchased, ensures supreme and uniform quality.

The word 'TABLOID' on your prescription secures for your patient a product of the same composition and activity always and everywhere.

The word 'TABLOID' may not be used except with reference to a Burroughs Wellcome & Co. product.

BEWARE OF IMITATIONS

Offenders are prosecuted rigorously in the interest of prescribers, dispensers, patients and the owners of the Trade Mark

The 'Sopwith Tabloid' becomes 'Sopwith Scout' after a complaint from Burroughs Wellcome.

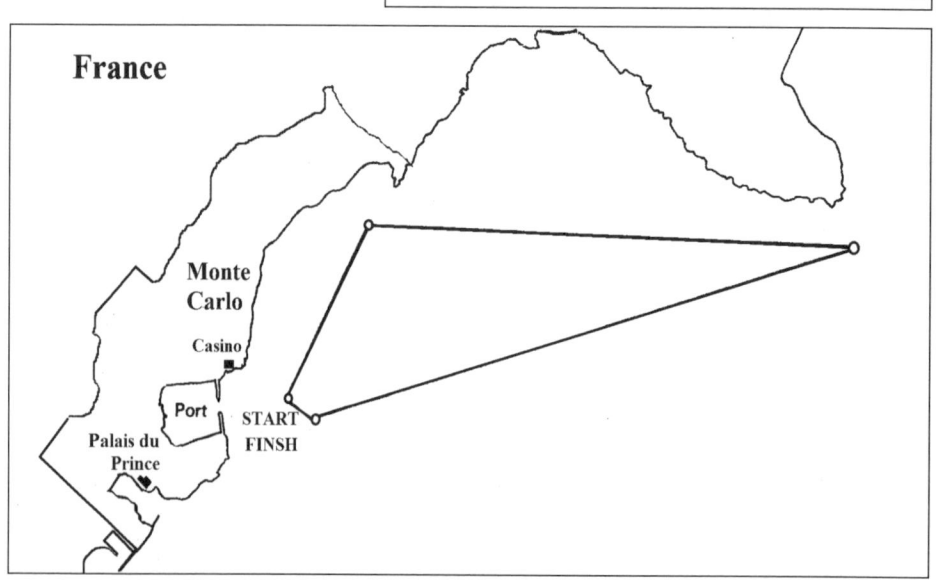

FAMOUS AIRMEN AND THEIR EQUIPMENTS

PIXTON

C. Howard Pixton won the Manville and Brooklands Aggregate Prizes, October 4, 1911, piloting a Bristol biplane with E.N.V. motor and an Avro biplane with a Green motor. In the picture he is seen to have a 'Tabloid' First-Aid Outfit strapped to his Avro biplane.

BURROUGHS WELLCOME & CO., LONDON

France

Monte Carlo
Casino
Port
START / FINISH
Palais du Prince

The Schneider Trophy Race, 1914.
A speed race, 150 nautical mile course, 28 laps.
At last, air supremacy for Britain!

HOWARD PIXTON
Pioneer Motorcylist
Pioneer Motorist
Pioneer Aviator
Flying Instructor
Competitor
Test pilot
Licence No. 50

C. Howard Pixton.

IN LOVING MEMORY
OF
SAMUEL FRANKLIN CODY

Who fell asleep, August 7th, 1913,

AGED 52 YEARS.

LIFE'S WORK WELL DONE,
LIFE'S RACE WELL RUN,
LIFE'S CROWN WELL WON—
NOW COMES REST.

INTERRED IN MILITARY CEMETERY, ALDERSHOT.

S. F. Cody

Principal Events Mentioned

Page	Event
1	Year of the Motor Car
2	Emancipation Day for Motorists
3	British Car Registration
3	First Flying Machine
8	First British Aero Show, Olympia London
9	English Channel Crossed
11	First Large Flying Meeting, Reims, France
11	Inauguration Gordon Bennett Contest
11	Aeronautical Exhibition, Frankfurt, Germany
12	French Flying Demonstration, Brooklands
13	France Leads the World
15	British Flying Licences Issued
15	London to Manchester Flight
28	First Fatality in Britain
31	Blackpool Air Meeting
35	Harvard University Flying Meeting, Boston, USA
39	Paris to London Flown
39	Irish Sea Crossed
40	First Mid-air Collision
50	Baron de Forest Contest
51	Michelin Tyre Contest
55	Formation, Air Battalion Beginning of Hendon
61	Brooklands to Brighton Race
61	Flying Demonstration, Haywards Heath
62	Manville Passenger Carrying Contest
67	Parliamentary Demonstration, Hendon
70	First Pilots' Strike
80	First Round Europe Race
80	First Round Britain Race
100	Pioneering Year of the Water Plane
112	Planned, Military Aeroplane Trials

Page	Event
113	In Paris
113	3rd International Paris Air Show
115	In Spain
116	In Germany
119	Formation, Royal Flying Corps, (RFC)
121	Germany again
123	Preparations, Military Trials
131	Results, Military Trials
134	Monoplane Ban, RFC
134	In Roumania
139	In Italy
140	Spain again
146	Round Britain Seaplane Race
148	French Stunt Demonstration, Brooklands
152	First Personal Accident Insurance for airmen
157	First Sopwith Tabloid biplane
159	Brooklands Handicap Race
161	International Schneider Trophy Seaplane Race, Monte Carlo, Monaco
172	Centre of Aviation shifts to Britain
176	Flying to the Danger of the Public
180	3rd Aerial Derby Race
184	How to recover from a Spin
186	War
188	The Aeroplane to be used in War
189	Formation, Aeronautical Inspection Department (AID)
190	First aerial attacks over Germany
191	First airship attacks over England
192	First Victoria Cross (VC) of the War awarded to a pilot
193	The Fokker Scourge
199	Formation, Royal Air Force (RAF)
201	War Ends

Principal Planes Mentioned

Year	Planes	Year	Planes
1903	The Wright Flyer	1913	Bristol Coanda biplanes
1906	Santos-Dumont Flying Machine	1913	Sopwith Bat Boats
1907	Blériot monoplane	1913	Sopwith Standards
1907	Voisin Box Kite	1914	Sopwith Seaplanes
1907	Farman Box Kite	1914	Sopwith Tabloids
1908	Roe No 1 pusher biplane	1914	Sopwith Schneider Tabloid Floatplane
1908	Cody British Army Plane No 1		
1910	Roe Triplanes	War	War Planes
1911	Roe Tractor biplane	1917	Leeds FE2b for India
1911	Bristol Box Kites	1918	AID DH9 for Lord Weir
1911	Bristol Prier monoplanes	1918	Handley Page 4-Engine biplane and 40 passengers
1912	Bristol Burney waterplane		
1912	Bristol Coanda monoplanes		

The Period Covered

Year	Page	Year	Page
1885	1	1910	15
1886	1	1911	52
1896	2	1912	111
1899	2	1913	140
1901	3	1914	156
1903	3	1915	191
1906	4	1916	193
1907	5	1917	195
1908	5	1918	199
1909	8		

Newspapers and Magazines

Daily Mail, The Star, Punch, The Car, The Graphic, Flight, Aeronautical Journal, The Aero, New York Herald, Boston Journal, The Sussex Daily News, The Westminster Gazette, The Aeroplane, The Observer, Pall Mall, Field, Clapham Express, The Standard, Belfast Weekly Telegraph, Western Daily Press, The Autocar, The Tatler, Melton Mowbray Times, Leeds Mercury, Yorkshire Herald, The Onlooker, London Opinion, Excelsior, Daily Telegraph, The Motor, The Surrey Herald, Daily Express, The Olympian.

Index

Aage, Prince, 181
Adams, Stanley, 72–3
Adby, Lady, 23–4
Albert, Prince, 172
Alcock, John, 54, 94, 151, 159, 182–4
Alexandra, Queen, 181
Alfonso XIII, King, 115, 140–1
Anstey, Rear Adml W.J., 202
Ashton, Lieut, 128
Asquith, H.H., 67
Astley, Henry, 22, 28, 52, 82, 86, 91, 134
Audemars, 83, 86, 88, 92

Babington, 190
Bagnall-Wild, Brig Gen, 190, 202
Baird, Maj, 200
Balfour, A.J., 67
Barber, 50, 67, 69, 105, 108
Barnes, George, 22, 30–1, 40, 50
Barnwell, Frank, 101
Barnwell, Harold, 201
Barnwell, Robert, 122, 148, 159–60, 182–3
Barrington-Kennett, Lieut, 98
Battenberg, Prince Leopold of, 141
Battenberg, Prince Maurice of, 191
Bayly, 189
Beatrice, Princess, 141
Beatty, Maj, 73, 193
Beaumont, 81–2, 85–6, 88, 91–3, 150, 166
Bedford, J.E., 197–8
Bell, Gordon, 49, 59, 130–1, 145, 177, 201
Bennett, Gordon, 11, 38, 40, 85, 161, 166, 172, 177, 186
Benson, 95, 145
Benz, 1
Beroniadi, 143
Bettington, Lieut Claude, 133
Bibesco, Prince, 135
Bielovucic, 124
Bier, 83, 92, 125
Billing, Eardley, 26–7

Billing, Mrs, 27, 187, 189
Billing, Pemberton, 148, 157, 177, 179
Birchenough, 182–3
Bjorkland, 182–3
Blackburn, Robert, 102, 138
Blacker, 73
Blanchet, 83, 86, 91
Blaschke, 125,
Blériot, Louis, 5–6, 8–11, 13, 16–17, 19, 22, 27, 30–1, 38–9, 42, 54–6, 62, 64, 67, 72–3, 82–3, 85, 92, 99–100, 104, 110, 114, 121, 124, 131, 134–5, 148–50, 159, 173, 180, 190, 203
Blériot, Madame, 10
Blondeau, Monsieur, 42, 50, 62, 95
Boris, Monsieur, 171
Boteanu, Gen, 135
Boyle, Alan, 22, 28, 30–1
Brewer, Griffith, 191
Briginshaw, 78–9, 102, 105
Briggs, 191
Brock, 182–3, 185
Brooke-Popham, Capt, 95, 126
Brookins, 37–8
Bulman, Maj G.P., 202
Burée, Lieut, 171
Burgess, 36
Burke, Capt, 98
Burney, Dennistoune, 101–102, 120, 147
Burney, Adml Sir Cecil, 101
Burri, 165–8, 171, 173
Busteed, Harry, 98–9, 114–17, 122–4, 129, 131–3, 140, 147, 154, 160

Calthrop, E.R., 200
Cammell, Lieut, 83, 86, 88, 91, 98, 100, 102, 145
Campbell, Lindsay, 122, 126, 145
Cantacuzène, Prince Serge, 122, 143
Carbery, Lord John, 165–7, 181–3, 185
Carr, 35–6, 182–3, 185
Carroll, Charles, 189
Cayley, Sir George, 103
Challenger, 94

Index

Chambenois, 125
Charteris, 124
Chavez, 34, 40
Chereau, Monsieur, 55, 75
Christescu, 143
Churchill, Winston, 67, 158, 203
Coanda, Henri, 114, 118, 123–5, 128–37, 139–40, 143–4, 147, 153, 187, 194, 203, 213
Cockburn, 11, 30, 100, 190
Cody, 6, 9, 12–13, 19, 28, 30, 46, 50–2, 67–9, 71, 83–4, 86–7, 91–3, 105–106, 108–109, 124–6, 129, 131–2, 136, 145–7, 174, 189–90, 201, 205, 213
Coles, Frank, 113, 116–17
Collett, 190
Collyer, 22, 43
Colmore, 31
Constantinescu, George, 194
Cox, Maj, 196
Curtiss, Glenn, 11, 14, 36, 38, 72, 100, 164–5, 186

Daimler, 1, 125
Davidson, Henry, 95
Davies, 42
Dawes, 31
de Bolotoff, Prince Serge, 152, 202
de Conneau, Lieut, 80–2, 150
de Forest, Baron, 50–2, 62, 178
de Havilland, Geoffrey, 127, 196, 202–203
de Môntalent, 91
des Moulinais, Brindejonc, 82, 86, 165
Delacombe, Harry, 99, 114, 169, 177, 187
Delacombe, Roy, 177
Delagrange, 5
Denman, Lord, 157
Desborough, Lord, 196–8
Desoutter, Marcel, 180
Destrem, Lieut, 171
Dickson, Capt Bertram, 28, 40–1, 54, 99, 139–40, 150
Douhet, Maj, 140
Drexel, 31, 34, 67–8
du Cros, Arthur, 67, 69
Ducrocq, Maurice, 50, 54, 148, 151

Dunne, 6, 108

Ebblewhite, 60, 62, 86
Edge, S.F., 23, 177–9
Edwards, 42
Egerton, 30
Elsdon, 159
Ena, Queen, 141
England, Gordon, 42–3, 50, 61, 65–7, 83–4, 86, 92, 96, 99, 102, 114, 122–3, 124, 127, 144, 147, 152, 177, 179
England, Geoff, 122, 143, 145
Esnault-Pelterie, Robert, 38, 150
Espanet, 165–8
Evans, W.H.B., 145, 147

Farman, Henry, 5–6, 9, 11, 15–17, 21–3, 27, 30, 38–40, 50, 62, 64, 68–9, 72–3, 75, 95–6, 100, 110, 114, 124–5, 135, 148, 150–1, 190, 193, 202–203
Farman, Maurice, 124–5, 151, 159, 203
Fellows, 144
Fenwick, 82, 86, 92, 124, 127, 131–2, 145
Feodorovna, Empress Marie, 181
Filitti, Maj, 135
Firth, J., 7
Fisher, 43, 50, 145
Flanders, Howard, 41
Fleming, 97
Fokker, 165, 193
Fulton, Col, 190, 192–3

Garros, Lieut Roland, 80, 164–6, 168–9, 171, 185, 193, 201
George V, King, 56, 69, 77, 132
George, David Lloyd, 67
Gibbs, Lieut, 27–8, 30
Gilmour, Graham, 21–3, 28–9, 43, 48, 50, 62, 64, 70, 83–5, 92, 95, 97, 99, 108, 118, 122
Gnosspelius, Oscar, 101
Goodden, 181
Gordon, Mrs Will, 41
Grace, Cecil, 15, 30, 32, 34, 50–1, 95
Gradisteau, Col, 135
Grahame-White, Claude, 15–17, 19, 21–4, 27–8, 30, 33–6, 38–40, 43, 50–1,

56, 67–70, 82, 84, 87, 109–10, 121, 150, 172, 177, 187, 202
Greswell, 42, 51, 62,
Grey, C.G., vi, 44, 49, 76, 85, 100, 111, 137, 145, 149, 152, 156, 162–3, 173, 177, 188, 191, 201
Grey, Lieut Spencer, 158, 177, 190

Haldane, Viscount, 67
Hall, Sam, 104
Hallam, Maude, 137
Halstead, 35, 39
Hamel, Gustav, 62, 64, 67–8, 83, 88, 91, 124, 150, 181–3, 186
Hamilton, Capt Patrick, 129, 133
Hammond, 57, 203
Handasyde, George 'Handy', 22, 43, 72, 118, 124, 131, 138, 144, 151, 186
Haralambie, Col, 135
Hardwick, Arkell, 138
Harford, Lieut, 105
Harrison, 154
Hartley, Sir William, 13
Hawker, Harry, 147, 154–9, 180, 183–5
Hawkins, 57–8, 203
Hearn, 185
Henderson, Brig Gen David, 95, 123, 145, 177–8, 196–8, 201
Hewetson, Maj, 144–5
Hewlett, Mrs, 42, 50, 62, 70–1, 95, 104–105
Hewlett, Maurice, 50
Holden, Col, 177–8
Holt, Thomas, 5, 177
Hotchkiss, Lieut Edward, 133
Houdini, Harry, 149
Hucks, B.C., 35–6, 83, 87, 91, 150, 177, 202
Hunt, 103

Islington, Lord, 197

Jacobs, Monsieur, 165
Jenkins, Conway, 42, 49, 61, 83, 87–8, 92, 199
John, Elton, viii,
Johnson, 37, 80
Jones, 34

Jullerot, 99, 114, 123, 144, 187

Kauper, 147, 154–5
Keith, 59
Kemp, 73–4, 82, 86, 92, 121
Kirk, Mildred, 35
Kitson, Mrs J.B., 194
Knight, 159

Laffont, 138
Lamson, Charles, 39
Lane, 22, 27
Latham, Hubert, 9–10, 72, 78, 123, 129
Lauder, Harry, 202
Lee, Arthur, 67, 69
Lee, Cedric, 152, 177, 186, 201
Lee, Zee Yee, 111
Lela, 125, 146
Levasseur, 165, 167–8, 171
Lincker, Gen Baron von, 116
Lloyd, Maj Lindsay, 12, 41, 70–2, 118, 177, 187
Loraine, Capt E.B., 123, 145
Loraine, Robert, 34, 39–40, 50–1, 67, 80, 83, 86, 92, 201
Lord, John, 32
Lovelace, Capt, 22
Low, 43, 50

MacDonald, Sidney, 35
Macfie, Robert, 28, 50, 52, 54, 113
Macintoon, Lieut, 117
Mackintosh, 102, 116–17
Macri, Maj, 135
Magnier, 11, 18
Mahl, Victor, 155–6, 162–7, 169–70, 179–80, 182, 184, 186, 201
Maitland, Capt Henry, 22, 40, 108, 202
Manville, Ernest, 62, 65, 78, 104–109, 173, 213–14
Mapplebeck, Lieut, 156
Marcus, Lord, 152
Marechal, 201
Marix, 190
Martin, J.V., 22, 35–6, 43, 56, 72, 118, 124, 131, 138–9, 144, 151, 159, 186, 189, 203
Massey, Capt, 98

Index

Maxim, Sir Hiram, 56
McArdle, 34
McClean, 50–1, 108, 147
McDonald, 124
McKenna, 67, 189
Merriam, Fred, 104, 151, 157, 159, 187, 192
Moineau, 124
Molesworth, 42, 57
Montagu, Lord, 5
Moore-Brabazon, H.P., 9, 12–13, 15, 29–30, 43, 50
Moris, Col, 139
Morison, Oscar, 22, 42, 50, 52, 61–2, 66–7, 80, 82, 84, 86, 92, 99, 108
Mosiant, 39

Napier, Gerald, 23, 94–5, 102, 145, 177
Neale, 22, 50
Negrescu, Lieut, 135, 143
Noel, 49, 106, 182–3, 185
Northcliffe, Lord, vi, 4, 9–10, 67, 81, 92–3, 147, 156

O'Gorman, Mervyn, 123
Ogden, 82
Ogilvie, Alec, 30, 38, 50–2, 85, 105, 108
Outram, Lieut Col H.W., 202
Oxley, Herbert, 35, 41, 47–9, 102–103, 138

Page, 82, 124, 131, 138, 148, 204
Paine, Capt Godfrey, 123
Parke, Lieut, 49, 73, 109, 124, 130, 138, 184–5
Parke, Rev A.W., 138
Parr, 124
Parvulescu, 135, 143
Pascanu, 143
Pashley, 42, 201
Passat, 56
Paterson, 82, 86, 92
Paulhan, Louis, 12, 17, 55, 150
Pégoud, Adolphe, 148–50, 193
Penfold, Capt, 122
Péquet, 42
Perreyon, 124
Perrin, Harold, 15, 17, 52, 85, 169, 177
Petitpierre, 73

Petre, 22, 43, 124, 138, 201
Pickles, Sydney, 122, 184
Pizey, Colin, 83–4, 86, 91, 97, 99, 140, 201
Platt, 7, 21–4, 31–3, 41, 43, 45
Pola, 138
Pooley, 42
Popovici, Andre, 143
Porte, Lieut, 86, 108, 186–7
Poynter, 42
Prade, Georges, 172
Pratt, Albert, 34
Preston, Harry, 31, 64, 82, 177
Prestwich, John A., 44
Prévost, 80, 127, 162, 165–6, 171
Prier, Monsieur, 56, 84–5, 99, 111–18, 121, 134, 140, 144, 153, 203
Prodger, Clifford, 204
Protopopescu, 135
Prowse, 59

Radeau, Col, 135
Radley, 28, 30–1, 38, 42, 83, 86, 92, 108, 119, 144, 147, 192
Rawlinson, 15, 30, 40
Raynham, 42, 73–4, 78, 105–106, 124, 130–1, 154, 184–5
Reid, George, 147
Renaux, 185
Reynolds, George, 60, 71, 83, 87, 92, 98, 201
Rhodes-Moorhouse, 125, 192
Richards, Tilghman, 152
Richy, Capt, 105
Robinson, Leefe, 194–5, 202
Roe, A.V., vi, 5–6, 9, 12–13, 18–77, 86, 96, 100–101, 107–108, 121, 134, 140, 154, 177, 186, 190, 192, 205
Roe, H.V., 18, 20, 22, 34, 45, 73, 192, 202, 213
Roempler, 159,
Rogers-Harrison, Lieut, 145
Rolls, C.S., 15, 28–30, 80, 95, 157, 203–204
Rose, 139
Royce, 29, 80, 157, 204
Rutherford, 158
Salmet, 146

Samson, 190
Santos-Dumont, 3–4, 19, 38, 114, 150
Sassoon, Ellis Victor, 42, 52, 76, 104, 201
Schneider, Jacques, vi–vii, 161–87, 193, 203, 215, 222
Schwann, Cdr, 72, 101
Seely, Col, 67, 119, 126, 128
Selfridge, Harry, 10, 202
Seti, 28, 43, 97, 122
Setti, Venkata Subba, 73
Short, 9, 12–13, 30–2
Sigrist, Fred, 154, 157
Sims, 35, 41
Sinclair, Archibald, 152
Singer, 30, 41, 155
Sippe, 73–4, 124, 129, 131, 140, 190
Skene, 151
Smith, E.V., 42, 95, 104
Smith, Sydney, 43
Snowden-Smith, 43, 62, 64
Sommer, Roger, 150
Sopwith, Tommy, 42–3, 46–7, 50–2, 56, 109–10, 118, 124, 131, 147, 153–87, 189–90, 192, 203, 222
Spencer, 41
Spooner, Stanley, 8, 177
Spottiswoode, John Herbert, 41, 79, 177, 202
Stirling, 41
Stoeffler, 165
Stokes, Mrs, 184
Stopes, Marie, 202
Strange, 182–3
Stutt, 154,
Sutcliffe, 7, 11
Sykes, Maj, 123

Tabuteau, 80, 83–4, 86, 92, 94, 99
Taft, President, 38
Taylor, 122
Teodoresco, Col, 135
Tetard, 34, 80, 99
Thaw, 165, 171
Thaye, Lieut, 194
Thomas, 40–1, 177
Thornly, J.B., 150
Thurston, Farnell, 102, 112–13, 115–16, 134, 137, 139–40

Togo, 122
Trenchard, Sir Hugh, 199
Tullibardine, Marquis of, 176–7, 186
Turner, C.C., 35, 76, 177

Valentine, James 'Jimmy', 42, 52, 62, 80, 83, 86, 88, 91–3, 99, 113, 115–16, 123–4, 127–8, 177, 201
Védrines, 80–1, 83, 86–8, 91–3, 124
Verne, Jules, 1
Verrier, 124, 182–3
Vickers, 73–4, 124, 131, 142, 148, 151, 159, 175, 181–2, 186–7, 198–9, 201, 203
Victoria, Queen, 1, 141
Vidart, 80
Vlaicu, Aurel, 136, 146
Voisin, Charles, 5, 8–9, 11, 21–2, 40, 150

Wakefield, E.W., 72, 91, 100–101
Warneford, Reginald, 192–3
Warren, Frederick, 104
Waterfall, Vincent, 151, 159, 189
Watkins, 42
Watney, Gordon, 72
Weir, Lord, 202–203
Wells, H.G., 6, 191
Weymann, 80, 83, 85, 87, 91, 165–6, 171
Whisky, 97, 122
Whistler, Mrs, 100
White, Sir George, 75, 82, 142, 201
White, Samuel, 142
White, Stanley, 75, 114, 120, 153
White Smith, 139–40
Wiasemsky, Princess, 152
Wickham, 42
Wijnmaalen, 83, 92
Wilhelm, Kaiser, 116, 186
Wilson, Sgt, 123, 145
Windham, Walter, 57, 72
Wolff, 108–109
Wood, Capt, 43, 73–4, 86, 175, 187, 202
Wright, Orville and Wilbur, 3–6, 8–14, 19, 37–9, 54, 108–10, 122, 172, 191
Wright, Frank, 176
Wyness-Stuart, Lieut, 133

Zeppelin, Count von, 12, 195
Zephyr, 172